The Bird

Books by Sheila Buff

The Birdfeeder's Handbook
Birding for Beginners
Corn Cookery

THE BIRDER'S SOURCEBOOK

Sheila Buff

Distributed by:
Airlife Publishing Ltd.
101 Longden Road, Shrewsbury SY3 9EB, England

Copyright © 1994 by Sheila Buff

ALL RIGHTS RESERVED. No part of this book may be reproduced in any manner without the express written consent of the publisher, except in the case of brief excerpts in critical reviews and articles. All inquiries should be addressed to: Lyons & Burford, 31 West 21 Street, New York, NY 10010.

Printed in the United States of America

Design by Laura Joyce Shaw

10 9 8 7 6 5 4 3 2 1

Buff, Sheila.
    The birder's sourcebook / Sheila Buff.
      p.   cm.
    Includes bibliographical references.
    ISBN 1-55821-278-7
    1. Bird watching—United States. 2. Bird watching—Canada. 3. Bird watching—United States—Directories. 4. Bird watching—Canada—Directories. 5. Bird watching. I. Title.
QL682.B84   1994
598'.07'2347—dc20   93-50202
                                                                       CIP

CONTENTS

Introduction ix

CHAPTER 1: WHERE TO BIRD

- United States National Parks 1
- United States National Forests 11
- United States National Wildlife Refuges 21
- Bureau of Land Management Sites 38
- Canadian National Parks 40
- Environment Canada Information Centers 45
- Canadian Wildlife Service Regional Offices 46
- Canadian Wildlife Service Natural Areas 46
- Canadian Provincial Agencies 49
- The Nature Conservancy Preserves 51
- National Audubon Society Sites 68

CHAPTER 2: TRAVEL INFORMATION FOR BIRDERS

- State Wildlife Offices 75
- Travel and Tourism Offices 81
  - United States 81
  - Canada 89
  - International 90

CHAPTER 3: BIRDERS ON TOUR

- Tour Operators 99
  - United States 99

        Canada   106
        North America   108
        International   109
Birding by Boat   114
Birding Events   120
        United States   121
        Canada   123
        International   123

CHAPTER 4: ORGANIZATIONS FOR BIRDERS

The Audubon Movement   124
        National Audubon Society Offices   125
        National Audubon Society
        Local Chapters   126
        Independent Audubon Societies   127
The American Birding Association   129
Cornell Laboratory of Ornithology   129
Nationwide Organizations   130
State Organizations   136
Canadian Organizations   138
International Organizations   140

CHAPTER 5: OPTICS FOR BIRDERS

Optics Manufacturers   143
Optics Retailers   145
Optics Repair   146
Optics Accessories   146
Bird Sound Amplification Devices   147

CHAPTER 6: THE EDUCATED BIRDER

Courses and Other Programs   148
        Home Study Course   148
        National Audubon Society
        Education Centers   149

National Audubon Society
Intern Program   149

Courses and Seminars   150

Volunteer Opportunities   151

Summer Youth Programs   152

Bird Photography Workshops   153

Zoos and Aviaries   154

    United States   154

    Canada   159

Museums with Significant
Ornithological Collections   159

CHAPTER 7: BOOKS, SOFTWARE, AND BEYOND

Books for Birders   162

Ornithology Libraries   165

    United States   166

    Canada   171

    England   172

Periodicals   172

Bird Slides   176

Birding Gear   176

Birding On-Line   178

APPENDICES

Appendix A: American Birding
Association Code of Ethics   181

Appendix B: Birding Hotlines   183

    United States   183

    Canada   186

Appendix C: State Birds   188

INTRODUCTION

Where are good places to bird? Who offers birding tours? What organizations are of interest to birders? Where can you buy birding paraphernalia? What libraries have special ornithology collections? In the course of writing several books about birds and birding, I accumulated vast quantities of brochures, catalogues, books, booklets, birding magazines, and scribbled notes. I found myself on the mailing list of seemingly every bird-related organization and company.

Free-lance writers are by nature pack rats and of necessity frugal. It seemed terribly wasteful to pile up all this information and not do something productive (and perhaps even profitable) with it. From the standpoint of a serious birdwatcher, the most useful thing was simply to compile it in an organized fashion. The result is *The Birder's Sourcebook*, a gathering of scattered information into one volume.

The information in *The Birder's Sourcebook* is as accurate and up-to-date as possible. Addresses and phone numbers do change, however. Generally speaking, mail sent to an old address is forwarded to the new, and a call to directory assistance will usually yield a new phone number. More difficult is the overall problem of keeping up with the dynamic world of birding. A few years ago, for example, a small handful of companies offered listing software. Today, a largish handful offer not only listing software but other types of useful programs as well. In addition, the number of computerized bulletin boards and on-line services has grown significantly. As birding continues to grow in popularity, additional services and suppliers are sure to appear.

A careful effort has been made to include as many places, organizations, services, manufacturers, retailers, and others completely and accurately as possible. Inclusion in this book is not an endorsement of any sort by the author or the publisher. Likewise, exclusion from this book is not deliberate in any way by the author or the publisher and does not imply criticism of any sort.

Corrections, updates, and additions for future editions of this book are encouraged. Please send them to the author in care of the publisher.

The Birder's Sourcebook

Chapter 1

WHERE TO BIRD

North Americans are extremely fortunate to have vast areas of public land on which to bird. The national parks of the United States and Canada alone cover millions of acres; national forests, wildlife refuges, and other federal lands cover millions more. The Nature Conservancy (a private, nonprofit organization) manages over 1,300 important natural sites in North America, while the National Audubon Society maintains an extensive system of preserves. In addition, parks, refuges, and other public lands in individual states and provinces are very extensive.

This chapter provides information about public lands on the national level in the United States and Canada. State and provincial lands are too numerous to list here, but see Chapter 2 to learn how to learn more.

UNITED STATES NATIONAL PARKS

The National Park system of the United States began with the establishment of Yellowstone National Park in 1872. Today the system comprises 357 areas covering more than 80 million acres in 49 states, the District of Columbia, American Samoa, Guam, Puerto Rico, Saipan, and the Virgin Islands. The parks are administered by the National Park Service, which is part of the U.S. Department of the Interior.

The National Park system encompasses a wide variety of parks, including many that are historical or commemorative. Most of the parks in the system, however, protect substantial natural areas. A national park contains a variety of resources and encompasses large land or water areas to help provide adequate protection of the resources. A national monument preserves a least one nationally significant resource, but is smaller and usually less diverse than a national park. National preserves protect natural resources, but some commercial extractive activities are sometimes allowed. Big Cypress and Big Thicket were the first national preserves, established in 1974. Preserving shoreline areas and off-shore islands are national lakeshores and national seashores. National rivers and wild and scenic riverways preserve ribbons of land bordering on free-flowing streams that have not been dammed or otherwise altered by humans. National scenic trails are generally long-distance footpaths winding through areas of great natural beauty.

Within many units of the National Park system are designated wilderness areas. These areas are managed to retain their wilderness character; no structures, motor vehicles, or roads are permitted.

The parks listed below are primarily national parks and monuments designed to protect natural areas. These sites are usually good places to watch birds. It should be noted, however, that many other easily accessible smaller sites within the National Park system (those that preserve Civil War battlefields, for instance) encompass hundreds of more or less undeveloped acres and are also good places to see birds.

Particularly during the popular summer months, plan your visit to a national park in advance. Reservations are often needed for campgrounds. To avoid crowds, try visiting popular areas during the off season, or try visiting some of the lesser-known areas. Some units of the system have limited or no federal facilities; access to these units may be restricted. In some cases, permits may be required for access. When contacting a national park, always ask for any current brochures, maps, bird lists, and other publications. These excellent publications are usually free or very inexpensive.

National Park Service Regional Offices

*North Atlantic Region
(CT, ME, MA, NH, NJ, NY, RI, VT)*

National Park Service
15 State Street
Boston, MA 02109

Mid-Atlantic Region (DE, MD, PA, VA, WV)

National Park Service
143 South Third Street
Philadelphia, PA 19106

National Capital Region (DC)

National Park Service
1100 Ohio Drive SW
Washington, DC 20242

*Southeast Region
(AL, FL, GA, KY, MS, NC, PR, SC, TN, VI)*

National Park Service
Richard B. Russell Federal Building
75 Spring Street SW
Atlanta, GA 30303

*Midwest Region
(IL, IN, IA, KS, MI, MN, MO, NE, OH, WI)*

National Park Service
1709 Jackson Street
Omaha, NE 68102

*Rocky Mountain Region
(CO, MT, ND, SD, UT, WY)*

National Park Service
Box 25287
Denver, CO 80225

*Southwest Region
(AZ, AR, LA, NM, OK, TX)*

National Park Service
Box 728
Santa Fe, NM 87504

*Western Region
(AZ, CA, HI, NV)*

National Park Service
600 Harrison Street, Suite 600
San Francisco, CA 94107

*Pacific Northwest Region
(ID, OR, WA)*

National Park Service
83 South King Street, Suite 212
Seattle, WA 98104

Alaska Region

National Park Service
2525 Gambell Street
Anchorage, AK 99503
(907) 261-2690

Alabama

Russell Cave National Monument
Route 1, Box 175
Bridgeport, AL 35740
(205) 495-2672

Alaska

Alagnak Wild River
c/o Katmai National Park
and Preserve
Box 7
King Salmon, AK 99613
(907) 246-3305

Aniakchak National
Monument and Preserve
Box 7
King Salmon, AK 99613
(907) 246-3305

Bering Land Bridge
National Park
Box 220
Nome, AK 99762
(907) 443-2522

Denali National Park
and Preserve
Box 9
Denali National Park, AK 99755
(907) 683-2294

Gates of the Arctic National
Park and Preserve
Box 74680
201 First Avenue
Fairbanks, AK 99707
(907) 456-0281

Glacier Bay National Park
and Preserve
Box 140
Bartlett Cove
Gustavus, AK 99826
(907) 697-2230

Katmai National Park
and Preserve
Box 7
King Salmon, AK 99613
(907) 246-3305

Kenai Fjords National Park
Box 1727
Seward, AK 99664
(907) 224-3874

Kobuk Valley National Park
Box 1029
Kotzebue, AK 99752
(907) 442-3890

Lake Clark National Park
and Preserve
4230 University Drive,
Suite 311
Anchorage, AK 99508
(907) 271-3751

Noatak National Preserve
Box 287
Kotzebue, AK 99752
(907) 442-3890

Wrangell-St. Elias National
Park and Preserve
Box 29
Glennallen, AK 99588
(907) 822-5234

Yukon-Charley Rivers
National Preserve
Box 167
Eagle, AK 99738
(907) 547-2233

National Park Service Regional Offices

American Samoa

The National Park of
American Samoa
Pago Pago
American Samoa 96799
(011) 684-633-70

Arizona

Chiricahua National
Monument
Dos Cabezas Route, Box 6500
Wilcox, AZ 85643
(602) 834-3560

Grand Canyon National Park
Box 129
Grand Canyon, AZ 86023
(602) 638-7888

Organ Pipe Cactus National
Monument
Route 1, Box 100
Ajo, AZ 85321
(602) 387-6849

Petrified Forest National Park
Petrified Forest National Park,
AZ 86028
(602) 524-6228

Saguaro National Monument
36933 Old Spanish Trail
Tucson, AZ 85730
(602) 670-6680

Arkansas

Buffalo National River
Box 1173
Harrison, AR 72601
(501) 741-5443

California

Cabrillo National Monument
Box 6670
San Diego, CA 92106
(619) 557-5450

Channel Islands National Park
1901 Spinnaker Drive
Ventura, CA 93001
(805) 644-8262

Golden Gate National
Recreation Area
Fort Mason, Building 201
San Francisco, CA 94123
(415) 556-0560

Joshua Tree National
Monument
74485 National Monument
Drive
Twentynine Palms, CA 92277
(619) 367-7511

Kings Canyon National Park
Three Rivers, CA 93271
(209) 565-3341

Muir Woods National
Monument
Mill Valley, CA 94941
(415) 388-2595

Pinnacles National Monument
Paicines, CA 95043
(408) 389-4485

Point Reyes National Seashore
Point Reyes, CA 94956
(415) 663-8522

Redwood National Park
1111 Second Street
Crescent City, CA 95531
(707) 464-6101

Sequoia National Park
Three Rivers, CA 93271
(209) 565-3341

Yosemite National Park
Box 577
Yosemite National Park, CA
95389
(209) 372-0200

Colorado

Rocky Mountain National
Park
Estes Park, CO 80517
(303) 586-2371 or
(303) 627-3471

District of Columbia

Rock Creek Park
5000 Glover Road NW
Washington, DC 20015
(202) 426-6833

Florida

Big Cypress National Preserve
Star Route, Box 110
Ochopee, FL 33943
(813) 695-2000

Biscayne National Park
Box 1369
Homestead, FL 33090
(305) 867-0634

Canaveral National Seashore
Box 6447
Titusville, FL 32782
(305) 867-0634

Everglades National Park
Box 279
Homestead, FL 33030
(305) 247-6211

Fort Jefferson National
Monument
c/o Everglades National Park
Box 279
Homestead, FL 33030
(305) 247-6211

Gulf Islands National Seashore
1801 Gulf Breeze Parkway
Gulf Breeze, FL 32561
(904) 934-2600

Timucuan Ecological and
Historic Preserve
c/o Fort Caroline National
Memorial
12713 Fort Caroline Road
Jacksonville, FL 32225
(904) 221-5568

Georgia

Cumberland Island National
Seashore
Box 806
St. Marys, GA 31558
(912) 882-4336

National Park Service Regional Offices

Hawaii

Haleakala National Park
Box 369
Makawao, HI 96768
(808) 572-9306

Indiana

Indiana Dunes National
Lakeshore
1100 North Mineral Springs
Road
Porter, IN 46304
(219) 926-7561

Kentucky

Mammoth Cave National Park
Mammoth Cave, KY 42259
(502) 758-2251

Louisiana

Jean Lafitte National Historic
Park and Preserve
423 Canal Street, Room 210
New Orleans, LA 70130
(504) 589-3882

Maine

Acadia National Park
Box 177
Bar Harbor, ME 04609
(207) 288-3338

Maryland

Assateague Island National
Seashore
Route 2, Box 294
Berlin, MD 21811
(301) 641-1441

Catoctin Mountain Park
6602 Foxville Road
Thurmont, MD 21788
(301) 663-9343

Chesapeake and Ohio Canal
National Historical Park
Box 4
Sharpsburg, MD 21782
(301) 739-4200

Massachusetts

Cape Cod National Seashore
South Wellfleet, MA 02663
(508) 349-3785

Michigan

Isle Royale National Park
87 North Ripley Street
Houghton, MI 49931
(906) 482-0986

Pictured Rocks
National Lakeshore
Box 40
Munising, MI 49862
(906) 387-3700

Sleeping Bear Dunes
National Lakeshore
Box 277
9922 Front Street
Empire, MI 49630
(616) 326-5134

Minnesota

Voyageurs National Park
HCR 9, Box 600
International Falls, MN 56649
(218) 283-9821

Mississippi

Gulf Islands National Seashore
3500 Park Road
Ocean Springs, MS 39564
(904) 932-5302

Missouri

Ozark National Scenic
Riverways
Box 490
Van Buren, MO 63965
(314) 323-4236

Montana

Bighorn Canyon National
Recreation Area
Box 458
Fort Smith, MT 59035
(406) 666-2412

Glacier National Park
West Glacier, MT 59936
(406) 888-5441

Nebraska

Missouri National
Recreation River
c/o Midwest Region,
National Park Service
1709 Jackson Street
Omaha, NE 68101
(402) 221-3481

Nevada

Great Basin National Park
Baker, NV 89311
(702) 234-7331

New Jersey

Gateway National
Recreation Area
Sandy Hook Unit
Box 437
Highland, NJ 07732
(201) 872-0115

New Mexico

White Sands National
Monument
Box 458
Alamogordo, NM 88310
(505) 437-1058

New York

Fire Island National Seashore
120 Laurel Street
Patchogue, NY 11772
(516) 289-4810

Gateway National
Recreation Area
Floyd Bennett Field, Building 69
Brooklyn, NY 11234
(718) 630-0126

North Carolina

Cape Hatteras
National Seashore
Route 1, Box 675
Manteo, NC 27954
(919) 473-2111

Cape Lookout National
Seashore
3601 Bridges Street, Suite F
Morehead City, NC 28557
(919) 240-1409

National Park Service Regional Offices

North Dakota

Theodore Roosevelt
National Park
Box 7
Medora, ND 58645
(701) 623-4466

Ohio

Cuyahoga Valley National
Recreation Area
15610 Vaughn Road
Brecksville, OH 44141
(216) 526-5256

Oregon

Crater Lake National Park
Box 7
Crater Lake, OR 97604
(503) 594-2211

Pennsylvania

Delaware Water Gap
National Recreation Area
Bushkill, PA 18324
(717) 588-6637

South Carolina

Congaree Swamp
National Monument
200 Caroline Sims Road
Hopkins, SC 29061
(803) 776-4396

South Dakota

Badlands National Park
Box 6
Interior, SD 57750
(605) 433-5361

Wind Cave National Park
Hot Springs, SD 57747
(605) 745-4600

Tennessee

Big South Fork National River
and Recreation Area
State Route 297
Oneida, TN 37841
(615) 879-4890

Great Smoky Mountains
National Park
Gatlinburg, TN 37738
(615) 436-1200

Obed Wild and Scenic River
Box 429
Wartburg, TN 37887
(615) 346-6294

Texas

Big Bend National Park
Big Bend National Park, TX
79834
(915) 477-2251

Big Thicket National Preserve
3785 Milam
Beaumont, TX 77701
(409) 839-2689

Guadalupe Mountains
National Park
HC 60, Box 400
Salt Flat, TX 79847
(915) 828-3251

WHERE TO BIRD

Rio Grande Wild
and Scenic River
c/o Big Bend National Park
Big Bend National Park, TX
79834
(915) 477-2251

Padre Island National Seashore
9405 South Padre Island Drive
Corpus Christi, TX 78418
(512) 937-2621

Virginia

Shenandoah National Park
Route 4, Box 348
Luray, VA 22835
(703) 999-2206

Virgin Islands

Buck Island Reef National
Monument
Box 160, Christiansted
St. Croix, VI 00820
(809) 773-1460

Virgin Islands National Park
6010 Estate Nazareth #10
St. Thomas, VI 00802
(809) 775-6238

Washington

Lake Chelan National
Recreation Area
2105 Highway 20
Sedro Woolley, WA 98284
(509) 682-4404

Mount Rainier National Park
Tahoma Woods, Star Route
Ashford, WA 98304
(206) 569-2211

North Cascades National Park
2105 Highway 20
Sedro Woolley, WA 98284
(206) 856-5700

Olympic National Park
600 East Park Avenue
Port Angeles, WA 98362
(206) 452-4501

West Virginia

New River Gorge
National River
Box 246
Glen Jean WV 25846
(304) 465-0508

Wisconsin

Apostle Island National
Lakeshore
Route 1, Box 4
Bayfield, WI 54814
(715) 779-3397

Wyoming

Grand Teton National Park
Drawer 170
Moose, WY 83012
(307) 733-2880

Yellowstone National Park
Box 168
Yellowstone National Park,
WY 82190
(307) 344-7381

UNITED STATES NATIONAL FORESTS

The 155 National Forests of the United States cover in total an area about the size of California, Oregon, and Washington states combined. More than 100,000 miles of trails wind through the forests. The National Forests are administered by the Forest Service of the U.S. Department of Agriculture.

Regional Offices

Forest Service Alaska
Region (AK)
Federal Office Building
709 West Ninth Street
Juneau, AK 99802
(907) 586-8863

Forest Service Eastern Region
(IL, IN, ME, MI, MN, MO,
NH, OH, PA, VT, WV, WI)
310 West Wisconsin Avenue,
Room 500
Milwaukee, WI 53203
(414) 297-3693

Forest Service Intermountain
Region (ID, NV, UT, WY)
Federal Building
324 25th Street
Ogden, UT 84401
(801) 625-5354

Forest Service Northern Region
(ID, MT)
Federal Building
200 East Broadway Street
Missoula, MT 59807
(406) 329-3511

Forest Service Pacific
Northwest Region (OR, WA)
319 SW Pine Street
Portland, OR 97208
(503) 221-2877

Forest Service Pacific
Southwest Region (CA, HI)
630 Sansome Street
San Francisco, CA 94111
(415) 705-2874

Forest Service Rocky Mountain
Region (CO, NE, SD, WY)
11177 West Eighth Avenue
Lakewood, CO 80225
(303) 236-9431

Forest Service Southern Region
(AL, AR, FL, GA, KY, LA, MS,
NC, PR, SC, TN, TX, VA, VI)
1720 Peachtree Road NW
Atlanta, GA 30367
(404) 347-4191

Forest Service Southwestern
Region (AZ, NM)
Federal Building
517 Gold Avenue SW
Albuquerque, NM 87102
(505) 842-3292

Alabama

William B. Bankhead,
Conecuh, Talladega, and
Tuskegee National Forests

WHERE TO BIRD

1765 Highland Avenue
Montgomery, AL 36107
(205) 832-4470

Alaska

Chugach National Forest
201 East Ninth Avenue,
Suite 206
Anchorage, AK 99501
(907) 271-2500

Tongass National Forest:
Chatham Area
204 Siginaka Way
Sitka, AK 99835
(907) 747-6671

Tongass National Forest:
Ketchikan Area
Federal Building
648 Mission
Ketchikan, AK 99833
(907) 225-3101

Tongass National Forest:
Stikine Area
201 12th Street
Petersburg, AK 99833
(907) 772-3841

Arizona

Apache–Sitgreaves
National Forest
Federal Building
309 South Mountain Avenue
Springerville, AZ 85938
(602) 333-4301

Coconino National Forest
2323 East Greenlaw Lane
Flagstaff, AZ 86004
(602) 556-7400

Coronado National Forest
300 West Congress Street,
6th Floor
Tucson, AZ 85701
(602) 670-6483

Kaibab National Forest
800 South Cortez
Prescott, AZ 86303
(602) 445-1762

Tonto National Forest
2324 East McDowell Road
Phoenix, AZ 85010
(602) 225-5200

Arkansas

Ouachita National Forest
Federal Building
100 Reserve Street
Hot Springs, AR 71902
(501) 321-5202

Ozark–St. Francis
National Forest
605 West Main Street
Russellville, AR 72801
(501) 968-2354

California

Angeles National Forest
701 North Santa Anita Avenue
Arcadia, CA 91006
(818) 574-1613

United States National Forests

Cleveland National Forest
10845 Rancho Bernardo Road,
Suite 200
Rancho Bernardo, CA 92127
(619) 673-6180

Eldorado National Forest
100 Forni Road
Placerville, CA 95667
(916) 644-6048

Inyo National Forest
873 North Main Street
Bishop, CA 95667
(619) 873-5841

Klamath National Forest
1312 Fairland Road
Yreka, CA 96097
(916) 842-6131

Lake Tahoe Basin
Management Unit
870 Emerald Bay Road, Suite 1
South Lake Tahoe, CA 96150
(916) 573-2600

Lassen National Forest
55 South Sacramento Street
Susanville, CA 96130
(916) 257-2151

Los Padres National Forest
6144 Calle Real
Goleta, CA 93117
(805) 683-6711

Mendocino National Forest
420 East Laurel Street
Willows, CA 95988
(916) 934-3316

Modoc National Forest
441 North Main Street
Alturas, CA 96101
(916) 233-5811

Plumas National Forest
159 Lawrence Street
Quincy, CA 95971
(916) 283-2050

San Bernardino
National Forest
1824 South Commercenter
Circle
San Bernardino, CA 92408
(714) 383-5588

Sequoia National Forest
900 West Grand Avenue
Porterville, CA 93257
(209) 784-1500

Shasta–Trinity National Forests
2400 Washington Avenue
Redding, CA 96001
(916) 246-5222

Sierra National Forest
1600 Tollhouse Road
Clovis, CA 93612
(209) 487-5155

Six Rivers National Forest
500 Fifth Street
Eureka, CA 95501
(707) 442-1721

Stanislaus National Forest
19777 Greenley Road
Sonora, CA 95370
(209) 532-3671

Tahoe National Forest
631 Coyote Street
Nevada City, CA 95959
(916) 265-4531

Colorado

Arapaho and Roosevelt
National Forests
240 West Prospect Road
Fort Collins, CO 80526
(303) 498-1100

Grand Mesa, Uncompahgre,
and Gunnison National Forests
2250 U.S. Highway 50
Delta, CO 81416
(303) 874-7691

Pike and San Isabel
National Forests
1920 Valley Drive
Pueblo, CO 81008
(719) 545-8737

Rio Grande National Forest
1803 West U.S. Highway 160
Monte Vista, CO 81144
(719) 852-5941

Routt National Forest
29587 West U.S. Highway 40,
Suite 20
Steamboat Springs, CO 80487
(303) 879-1722

San Juan National Forest
701 Camino Del Rio,
Room 301
Durango, CO 81301
(303) 247-4874

White River National Forest
Old Federal Building
Box 948
Glenwood Springs, CO 81602
(303) 945-2521

Florida

Apalachicola, Ocala, and
Osceola National Forests
USDA Forest Service
227 North Bronough Street,
Suite 4061
Tallahassee, FL 32301
(904) 681-7265

Georgia

Chattahoochee-Oconee
National Forests
508 Oak Street NW
Gainesville, GA 30501
(404) 536-0541

Idaho

Boise National Forest
1750 Front Street
Boise, ID 83702
(208) 364-4100

Caribou National Forest
Federal Building, Suite 294
250 South Fourth Avenue
Pocatello, ID 83201
(208) 236-7500

Challis National Forest
U.S. Highway 93 North
Challis, ID 83226
(208) 879-2285

Clearwater National Forest
12730 Highway 12
Orofino, ID 83544
(208) 476-4541

Idaho Panhandle National
Forests: Coeur d'Alene,
Kaniksu, and
St. Joe National Forests
1201 Ironwood Drive
Coeur d'Alene, ID 83814
(208) 765-7223

Nez Perce National Forest
East U.S. Highway 13
Route 2, Box 475
Grangeville, ID 83530
(208) 983-1950

Payette National Forest
106 West Park Street
McCall, ID 83638
(208) 634-8151

Salmon National Forest
U.S. Highway 93 North
Salmon, ID 83467
(208) 726-2215

Sawtooth National Forest
2647 Kimberly Road East
Twin Falls, ID 83301
(208) 737-3200

Targhee National Forest
420 North Bridge Street
St. Anthony, ID 83445
(208) 624-3151

Illinois

Shawnee National Forest
901 South Commercial Street
Harrisburg, IL 62946
(618) 253-7114

Indiana

Wayne–Hoosier
National Forests
811 Constitution Avenue
Bedford, IN 47421
(812) 275-5987

Kentucky

Daniel Boone National Forest
100 Vaught Road
Winchester, KY 40391
(606) 745-3100

Louisiana

Kisatchie National Forest
2500 Shreveport Highway
Pineville, LA 71360
(318) 473-7160

Michigan

Hiawatha National Forest
2727 North Lincoln Road
Escanaba, MI 49829
(906) 786-4062

Huron–Manistee
National Forests
421 South Mitchell Street
Cadillac, MI 49601
(616) 775-2421

Ottawa National Forest
2100 East Cloverland Drive
Ironwood, MI 49938
(906) 932-1330

Minnesota

Chippewa National Forest
Route 3, Box 244
Cass Lake, MN 56633
(218) 335-2226

Superior National Forest
515 West First Street
Duluth, MN 55801
(218) 720-5324

Mississippi

Bienville, Delta, DeSoto, Holly Springs, Homochitto, and Tombigbee National Forests
100 West Capitol Street, Suite 1141
Jackson, MS 39269
(601) 965-4391

Missouri

Mark Twain National Forest
401 Fairgrounds Road
Rolla, MO 65401
(314) 364-4621

Montana

Beaverhead National Forest
420 Barrett Street
Dillon, MT
(406) 683-3900

Bitterroot National Forest
1801 North First Street
Hamilton, MT 59840
(406) 363-3131

Custer National Forest
2602 First Avenue South
Billings, MT 59840
(406) 657-6361

Deerlodge National Forest
Federal Building
Box 400
Butte, MT 59703
(406) 496-3400

Flathead National Forest
1935 Third Avenue East
Kalispell, MT 59901
(406) 755-5401

Gallatin National Forest
Federal Building
10 East Babcock Street
Bozeman, MT 59771
(406) 587-6701

Helena National Forest
Federal Building
301 South Park, Room 334
Helena, MT 59626
(406) 449-5201

Kootenai National Forest
506 U.S. Highway 2 West
Libby, MT 59923
(406) 293-6211

Lewis and Clark
National Forest
1101 15th Street North
Great Falls, MT 59403
(406) 791-7700

United States National Forests

Lolo National Forest
Building 24, Fort Missoula
Missoula, MT 59801
(406) 329-3750

Nebraska

Nebraska National Forest
270 Pine Street
Chadron, NE 69337
(308) 432-0300

New Hampshire

White Mountain
National Forest
Federal Building
719 North Main Street
Laconia, NH 03247
(603) 528-8721

New Mexico

Carson National Forest
Forest Service Building
208 Cruz Alta Road
Taos, NM 87571
(505) 758-6200

Cibola National Forest
2113 Osuna Road NE, Suite A
Albuquerque, NM 87713
(505) 761-4650

Gila National Forest
2610 North Silver Street
Silver City, NM 88061
(505) 388-8201

Lincoln National Forest
Federal Building
Alamogordo, NM 88310
(505) 437-6030

Santa Fe National Forest
Pinon Building
1220 St. Francis Drive
Santa Fe, NM 87504
(505) 988-6940

Nevada

Humboldt National Forest
976 Mountain City Highway
Elko, NV 89801
(702) 738-5171

Toiyabe National Forest
1200 Franklin Way
Sparks, NV 89431
(702) 355-5301

North Carolina

Croatan, Nantahala, Pisgah,
and Uwharrie National Forests
100 Otis Street
Asheville, NC 28802
(704) 257-4200

Oregon

Columbia River Gorge
National Scenic Area
902 Wasco Avenue
Hood River, OR 97031
(503) 386-2333

Deschutes National Forest
1645 U.S. Highway 20 East
Bend, OR 97701
(503) 388-2715

Fremont National Forest
524 North G Street
Lakeview, OR 97630
(503) 947-2151

Malheur National Forest
139 NE Dayton Street
John Day, OR 97845
(503) 575-1731

Mount Hood National Forest
2955 NW Division Street
Gresham, OR 97030
(503) 666-0700

Ochoco National Forest
3000 East Third
Prineville, OR 97754
(503) 447-6247

Rogue River National Forest
Federal Building
333 West Eighth Street
Medford, OR 97501
(503) 776-3600

Siskiyou National Forest
200 NE Greenfield Road
Grants Pass, OR 97526
(503) 479-5301

Siuslaw National Forest
4077 Research Way
Corvallis, OR 97339
(503) 750-7000

Umatilla National Forest
2517 SW Hailey Avenue
Pendleton, OR 97801
(503) 276-3811

Umpqua National Forest
2900 NW Stewart Parkway
Roseburg, OR 97470
(503) 672-6601

Wallowa–Whitman
National Forest
1550 Dewey Avenue
Baker City, OR 97814
(503) 523-6391

Willamette National Forest
211 East Seventh Avenue
Eugene, OR 97440
(503) 465-6521

Winema National Forest
2819 Dahlia Street
Klamath Falls, OR 97601
(503) 883-6714

Pennsylvania

Allegheny National Forest
Spiridon Building
222 Liberty Street
Warren, PA 16365
(814) 723-5150

Puerto Rico and Virgin Islands

Caribbean National Forest
University of Puerto Rico
Agricultural Experiment
Station
Call Box 25000
Rio Piedras, PR 00928
(809) 766-5335

South Carolina

Francis Marion–Sumter
National Forests
1835 Assembly Street,
Room 333
Columbia, SC 29201
(803) 765-5222

South Dakota

Black Hills National Forest
Highway 385 North
Custer, SD 57730
(605) 673-2251

Tennessee

Cherokee National Forest
2800 North Ocoee Street NW
Cleveland, TN 37320
(615) 476-9700

Texas

Angelina, Davy Crockett,
Sabine, and Sam Houston
National Forests
Homer Garrison
Federal Building
701 North First Street
Lufkin, TX 75901
(409) 639-8501

Utah

Ashley National Forest
355 North Vernal Avenue
Vernal, UT 84078
(801) 789-1181

Dixie National Forest
82 North 100 East
Cedar City, UT 84720
(801) 586-2421

Fishlake National Forest
115 East 900 North
Richfield, UT 84701
(801) 896-4491

Manti–LaSal National Forest
599 West Price River Drive
Price, UT 84501
(801) 637-2817

Uinta National Forest
88 West 100 North
Provo, UT 84601
(801) 377-5780

Wasatch–Cache
National Forest
8236 Federal Building
125 South State Street
Salt Lake City, UT 84138
(801) 524-5030

Vermont

Green Mountain and Finger
Lakes National Forests
Federal Building
151 West Street
Rutland, VT 05701
(802) 773-0300

Virginia

George Washington
National Forest
101 North Main Street
Harrison Plaza
Harrisonburg, VA 22801
(703) 433-2491

Jefferson National Forest
210 Franklin Road SW
Roanoake, VA 24001
(703) 982-6270

Washington

Colville National Forest
695 South Main Street
Colville, WA 99114
(509) 684-3711

Gifford Pinchot
National Forest
6926 East Fourth Plain
Boulevard
Vancouver, WA 98668
(206) 696-7500

Mt. Baker–Snoqualmie
National Forests
21905 64th Avenue West
Mountlake Terrace, WA 98043
(206) 775-9702

Okanogan National Forest
1240 South Second Avenue
Okanogan, WA 98840
(509) 826-3275

Olympic National Forest
1835 Black Lake Boulevard SW
Olympia, WA 98502
(206) 956-2300

Wenatchee National Forest
301 Yakima Street
Wenatchee, WA 98801
(509) 662-4335

West Virginia

Monongahela National Forest
USDA Building
200 Sycamore Street
Elkins, WV 26241
(304) 636-1800

Wisconsin

Chequamegon National Forest
1170 Fourth Avenue South
Park Falls, WI 54552
(715) 762-2461

Nicolet National Forest
Federal Building
68 South Stevens Street
Rhinelander, WI 54501
(715) 362-3415

Wyoming

Bighorn National Forest
1969 South Sheridan Avenue
Sheridan, WY 82801
(307) 672-0751

Bridger–Teton National Forest
Forest Service Building
340 North Cache
Jackson, WY 83001
(307) 733-2752

Medicine Bow National Forest
2468 Jackson Street
Laramie, WY 82070
(307) 745-8971

Shoshone National Forest
225 West Yellowstone Avenue
Cody, WY 82414
(307) 527-6241

UNITED STATES NATIONAL WILDLIFE REFUGES

The first national wildlife refuge in the United States was tiny Pelican Island in Florida, established under President Theodore Roosevelt in 1903. Since then, the system has grown to include nearly 450 refuges encompassing more than 90 million acres of land and water managed specifically for wildlife. Most but not all refuges are open to the public, however; the listing below includes only those that are. The addresses below are those of the offices administering the refuges, so they do not necessarily reflect the sites' location.

When planning a visit to a national wildlife refuge, contact the refuge office in advance to receive specific information regarding directions for access, special activities, regulations, and other helpful information. When you arrive at the refuge, check in with the visitor center or contact station to get up-to-date information. You can also get information about visiting national wildlife refuges from:

National Wildlife Refuge Association
10824 Fox Hunt Lane
Potomac, MD 20854

National Headquarters

U.S. Fish and Wildlife Service
4401 North Fairfax Drive, Room 130
Arlington, VA 22203

Regional Offices

Region 1 (CA, ID, HI, NV, OR, WA)
Eastside Federal Complex
911 NE 11th Avenue
Portland, OR 97232
(503) 231-6118

Region 2 (AZ, NM, OK, TX)
Box 1306
Albuquerque, NM 87103
(505) 766-2321

Region 3 (IL, IN, IA, MI, MN, MS, OH, WI)
Federal Building
Fort Snelling
Twin Cities, MN 55111
(612) 725-3563

Region 4 (AR, AL, FL, GA, KY, LA, MS, NC, SC, TN, PR)
Richard B. Russell Federal Building
75 Spring Street, SW
Atlanta, GA 30303
(404) 331-3588

Region 5 (CT, DE, ME, MD, MA, NH, NJ, NY, PA, RI, VT, WV)
One Gateway Center, Suite 700
Newton Corner, MA 02158
(617) 965-5100

Region 6 (CO, KS, MT, NE, ND, SD, UT, WY)
Denver Federal Center
Box 25486
Lakewood, CO 80228
(303) 236-7920

Region 7 (AK)
1011 East Tudor Road
Anchorage, AK 99503
(907) 786-3542

Alabama

Bon Secour
Box 1650
Gulf Shores, AL 36542
(205) 968-8623

Choctaw
Box 808
Jackson, AL 36545
(205) 246-3583

Eufaula (Alabama and Georgia)
Route 2, Box 97-B
Eufaula, AL 36027
(205) 687-4065

Wheeler
Box 1643
Decatur, AL 35602
(205) 353-7243

Alaska

Alaska Maritime (headquarters)
2355 Kachemak Bay Drive, Suite 101
Homer, AK 99603
(907) 235-6546
Includes Alaska Peninsula, Aleutian Islands, Bering Sea, Chukchi Sea, Gulf of Alaska units

Alaska Peninsula/Becharof
Box 277
King Salmon, AK 99613
(907) 246-3339

Arctic
101 12th Avenue
Box 20
Fairbanks, AK 99701
(907) 456-0250

Innoko
Box 69
McGrath, AK 99627
(907) 524-3251

Izembek
Box 127
Cold Bay, AK 99571
(907) 532-2445

Kanuti
101 12th Avenue
Box 20
Fairbanks, AK 99701
(907) 456-0329

Kenai
Box 2139
Soldotna, AK 99669
(907) 262-7021

United States National Wildlife Refuges

Kodiak
1390 Buskin River Road
Kodiak, AK 99615
(907) 487-2600

Koyukuk/Nowitna
Box 287
Galena, AK 99741
(907) 656-1231

Selawik
Box 270
Kotzebue, AK 99752
(907) 442-3799

Tetlin
Box 155
Tok, AK 99780
(907) 883-5312

Togiak
Box 270
Dillingham, AK 99576
(907) 842-1063

Yukon Delta
Box 346
Bethel, AK 99559
(907) 543-3151

Yukon Flats
101 12th Avenue
Box 20
Fairbanks, AK 99701
(907) 456-0407

Arizona

Buenos Aires
Box 106
Sasabe, AZ 85633
(602) 823-4251

Cabeza Prieta
Box 418
Ajo, AZ 85321
(602) 387-6483

Cibola (Arizona and
California)
Box AP
Blythe, CA 92225
(602) 857-3253

Havasu (Arizona
and California)
Box A
Needles, CA 92363
(619) 326-3853

Imperial (Arizona
and California)
Box 72217
Martinez Lake, AZ 85364
(602) 783-3371

Kofa
Box 6290
Yuma, AZ 85364
(602) 783-7861

San Bernadino
RR 1, Box 228R
Douglas, AZ 85607
(602) 364-2104

Arkansas

Felsenthal
Box 1157
Crossett, AR 71635
(501) 364-3167

Holla Bend
Box 1043
Northeast Arkansas Refuges
Box 279
Turrell, AR 72384
(501) 343-2595
Includes Big Lake, Cache River, Wapanocca, White River units

California

Cibola (*See* Arizona)

Havasu (*See* Arizona)

Hopper Mountain
Box 3817
Ventura, CA 93006
(805) 644-1766

Imperial (*See* Arizona)

Kern
Box 670
Delano, CA 93216
(805) 725-2767

Klamath Basin Refuges
Route 1, Box 74
Tulelake, CA 96134
(916) 667-2231
Includes Bear Valley, Clear Lake, Klamath Forest, Lower Klamath (California and Oregon), Tule Lake, Upper Klamath

Modoc
Box 1610
Alturas, CA 96101
(916) 233-3572

Sacramento Valley Refuges
Route 1, Box 311
Willows, CA 95988
(916) 934-2801
Includes Butte Sink, Colusa, Delevan, Sacramento River, Sutter, Willow Creek-Lurline units

Salton Sea/Coachella
Box 120
Calipatria, CA 92233
(619) 348-5278

San Francisco Bay
Box 524
Newark, CA 94560
(415) 792-0222
Includes Antioch Dunes, Castle Rock, Ellicott Slough, Farallon, Humboldt Bay, Salinas River, San Pablo Bay units

San Luis
Box 2176
Los Banos, CA 93635
(209) 826-3508
Includes Grasslands, Kesterson, Merced, San Joaquin River units

Southern California Coastal Complex
Box 335
Imperial Beach, CA 92030
(619) 575-1290

Colorado

Alamosa/Monte Vista
Box 1148
Alamosa, CO 81101
(719) 589-4021

United States National Wildlife Refuges

Arapaho
Box 457
Walden, CO 80480
(303) 723-8202

Browns Park
1318 Highway 318
Maybell, CO 81640
(303) 365-3613

Connecticut

Salt Meadow
Box 307
Charlestown, RI 02813
(401) 364-9124

Stewart B. McKinney
910 Lafayette Boulevard,
Room 210
Bridgeport, CT 06604
(203) 399-2513

Delaware

Bombay Hook
Route 1, Box 147
Smyrna, DE 19977
(302) 653-9345

Prime Hook
Route 1, Box 195
Milton, DE 19968
(302) 684-8419

Florida

Arthur R. Marshall
Loxahatchee
Route 1, Box 278
Boynton Beach, FL 33437
(407) 732-3684
Includes Hobe Sound unit

Chassahowitzka
Box 4139
Homosassa, FL 32647
(904) 382-2201
Includes Crystal River, Egmont Key, Passage Key, Pinellas units

Florida Panther
2629 South Horseshoe Drive
Naples, FL 33942
(813) 353-8442

J.N. "Ding" Darling
One Wildlife Drive
Sanibel, FL 33957
(813) 472-1100
Includes Caloosahatchee, Island Bay, Matlacha Pass, Pine Island units

Lake Woodruff
Box 488
DeLeon Springs, FL 32028
(904) 985-4673

Lower Suwanee/Cedar Keys
Route 1, Box 1193C
Chiefland, FL 32626
(904) 493-0238

Merritt Island
Box 6504
Titusville, FL 32780
(407) 867-0667
Includes Pelican Island and St. Johns units

National Key Deer
Box 510
Big Pine Key, FL 33043
(305) 872-2239
Includes Crocodile Lake, Great White Heron, Key West units

St. Marks
Box 68
St. Marks, FL 32355
(904) 925-6121

St. Vincent
Box 447
Apalachicola, FL 32320
(904) 653-8808

Georgia

Eufaula (*See* Alabama)

Savannah Coastal Refuges
Box 8487
Savannah, GA 31412
(912) 944-4415
Includes Blackbeard Island, Harris Neck, Pinckney Island, Savannah (Georgia and South Carolina), Tybee, Wassaw, Wolf Island

Okefenokee/Banks Lake
Route 2, Box 338
Folkston, GA 31537
(912) 496-7366

Piedmont/Bond Swamp
Route 1, Box 670
Round Oak, GA 31038
(912) 986-5441

Hawaii

Hawaiian and Pacific Islands Complex
Box 50167
300 Ala Moana Boulevard
Honolulu, HI 96850
(808) 541-1201
Includes Hakalau Forest, James C. Campbell, Kilauea Point, Remote Island Refuges, Johnston Atoll units

Oahu
Box 340
Haleiwa, HI 96712
(808) 637-6330

Idaho

Deer Flat
13751 Upper Embankment Road
Box 448
Nampa, ID 83653
(208) 467-9278

Kootenai
HCR 60, Box 283
Bonners Ferry, ID 83805
(208) 267-3888

Southeast Idaho Refuge Complex
1246 Yellowstone Avenue
Pocatello, ID 83201
(208) 237-6615
Includes Bear Lake, Camas, Gray's Lake, Minidoka units

United States National Wildlife Refuges

Illinois

Chautauqua/Meredosia
Route 2
Havana, IL 62644
(309) 535-2290

Crab Orchard
Box J
Carterville, IL 62918
(618) 997-3344

Mark Twain
311 North Fifth Street,
Suite 100
Quincy, IL 62301
(217) 224-8580
Includes Annada, Brussels,
Wapello districts

Upper Mississippi River (*See also* Minnesota)

Savanna District
Post Office Building
Savanna, IL 61074
(815) 273-2732

Indiana

Muscatatuck
Route 7, Box 189A
Seymour, IN 47274
(812) 522-4352

Patoka
Box 510
Winslow, IN 47598
(812) 789-2102

Iowa

DeSoto (Iowa and Nebraska)
Route 1, Box 114
Missouri Valley, IA 51555
(712) 642-4121

Union Slough
Route 1, Box 52
Titonka, IA 50480
(515) 928-2523

Upper Mississippi River (*See also* Minnesota)

McGregor District
Box 460
McGregor, IA 52157
(319) 873-3423

Walnut Creek
Box 399
Prairie City, IA 50228
(515) 994-2415

Kansas

Flint Hills
Box 128
Hartford, KS 66854
(316) 392-5553

Kirwin
Route 1, Box 103
Kirwin, KS 67644
(913) 543-6673

Quivira
Route 3, Box 48A
Stafford, KS 67578
(316) 486-2393

Louisiana

Bogue Chitto
1010 Gause Boulevard,
Building 936
Slidell, LA 70458
(504) 646-7555
Includes Bayou Savage and
Breton Delta units

Cameron Prairie
Route 1, Box 642
Bell City, LA 70630
(318) 598-2216

Catahoula
Drawer LL
Jena, LA 71342
(318) 992-5261

D'Arbonne/Upper Ouachita
Box 3065
Monroe, LA 71201
(318) 325-1735

Lacassine
Route 1, Box 186
Lake Arthur, LA 70549
(318) 774-5923
Includes Atchafalaya and Shells
Keys units

Lake Ophelia/Grand Cote
Box 256
Marksville, LA 71351
(318) 253-4131

Louisiana Wetlands
Management District
Box 1601
Monroe, LA 71210
(318) 726-4400

Sabine
MRH 107
Hackberry, LA 70645
(318) 762-3816

Tensas River
Route 2, Box 295
Tallulah, LA 71282
(318) 574-2664

Maine

Moosehorn
Box X
Calais, ME 04619
(207) 454-3521
Includes Cross Island, Franklin
Island, Seal Island units

Petit Manan
Box 279
Milbridge, ME 04658
(207) 546-2124

Rachel Carson
Route 2, Box 751
Wells, ME 04090
(207) 646-9226

Skunkhaze Meadows
1033 South Main Street
Old Town, ME 04468
(207) 581-3670

Maryland

Blackwater
Route 1, Box 121
Cambridge, MD 21613
(301) 228-2692
Includes Martin and
Susquehanna units

United States National Wildlife Refuges

Eastern Neck
Route 2, Box 225
Rock Hall, MD 21661
(410) 639-7056

Patuxent
Route 197
Laurel, MD 20708
(301) 498-0300

Massachusetts

Great Meadows
Weir Hill Road
Sudbury, MA 01776
(508) 443-4661
Includes John Hay (New Hampshire), Massasoit, Nantucket, Oxbow, Wapack (New Hampshire) units

Monomoy
Wikis Way, Morris Island
Chatham, MA 02633
(508) 945-0594

Parker River
Northern Boulevard, Plum Island
Newburyport, MA 01950
(508) 465-5753
Includes Pond Island (Maine), Rachel Carson, Thacher Island units

Michigan

Seney
Seney, MI 49883
(906) 586-9851
Includes Harbor Island, Huron units

Shiawassee
6975 Mower Road
Route 1
Saginaw, MI 48601
(517) 777-5930
Includes Michigan Islands, Wyandotte units

Minnesota

Agassiz
Middle River, MN 56737
(218) 449-4115

Big Stone
25 Northwest Second Street
Ortonville, MN 56278
(612) 839-3700

Minnesota Valley
4101 East 80th Street
Bloomington, MN 55420
(612) 854-5900

Minnesota Wetlands Complex
Route 1, Box 76
Fergus Falls, MN 56537
(218) 739-2291
Includes Detroit Lake, Fergus Falls, Litchfield, Mamden Slough, Morris, Windom units

Rice Lake/Mille Lacs
Route 2, Box 67
McGregor, MN 55760
(218) 768-2402

Sherburne
Route 2
Zimmerman, MN 55398
(612) 389-3323

WHERE TO BIRD

Tamarac
Rural Route
Rochert, MN 56578
(218) 847-2641

Upper Mississippi (includes
Illinois, Iowa, Minnesota,
Wisconsin)
51 East Fourth Street
Winona, MN 55987
(507) 452-4232
Includes Winona District

Mississippi

Mississippi Sandhill Crane
Box 699
Gautier, MS 39553
(601) 497-6322

Mississippi Wildlife
Management District
Box 1070
Grenada, MS 38901
(601) 226-8286

Noxubee
Route 1, Box 142
Brooksville, MS 39739
(601) 323-5548

St. Catherine Creek
Box 18639
Natchez, MS 39122
(601) 442-6696

Yazoo
Route 1, Box 286
Hollandale, MS 38748
(601) 839-2638
Includes Hillside, Morgan
Brake, Panther Swamp units

Missouri

Mingo/Pilot Knob
Route 1, Box 103
Puxico, MO 63960
(314) 222-3589

Squaw Creek
Box 101
Mound City, MO 64470
(816) 442-3187

Swan Lake
Box 68
Sumner, MO 64681
(816) 856-3323

Montana

Benton Lake
Box 450
Black Eagle, MT 59414
(406) 727-7400

Bowdoin
Box J
Malta, MT 59538
(406) 654-2863
Includes Black Coulee,
Creedman Coulee, Hewitt
Lake, Lake Thibadeau units

Charles M. Russell
Box 110
Lewistown, MT 59457
(406) 538-8706
Includes Hailstone, Halfbreed
Lake, Lake Mason, UL Bend,
War Horse units

United States National Wildlife Refuges

Lee Metcalf
Box 257
Stevensville, MT 59870
(406) 777-5552

Medicine Lake/Lamesteer
HC 51, Box 2
Medicine Lake, MT 59247
(406) 789-2305

National Bison Range
Moiese, MT 59824
(406) 644-2211
Includes Northwest Montana
Wetlands Management District

Red Rock Lakes
Monida Star Route, Box 15
Lima, MT 59739
(406) 276-3347

Nebraska

Crescent Lake
HC 68, Box 21
Ellsworth, NE 69340
(308) 762-4893
Includes North Platt unit

Fort Niobrara/Valentine
Hidden Timber Route
HC 14, Box 67
Valentine, NE 69201
(402) 376-3789

Rainwater Basin Wildlife
Management District
Box 1686
Kearney, NE 68848
(308) 236-5015

Nevada

Desert Refuge Complex
1500 North Decatur Boulevard
Las Vegas, NV 89108
(702) 646-3401
Includes Ash Meadows, Desert
National Wildlife Range,
Pahranagat units

Ruby Lake
Ruby Valley, NV 89833
(702) 779-2237

Stillwater
Box 1236
1510 Rio Vista Road
Fallon, NV 89406
(702) 423-5128
Includes Anaho Island,
Fallon units

New Hampshire
(*see* **Massachusetts**)

New Jersey

Barnegat
Box 544
Barnegat, NJ 08005
(609) 698-1387

Cape May
Box 72
Great Creek Road
Oceanville, NJ 08231
(609) 652-1665

Edwin B. Forsythe/Brigantine
Box 72
Oceanville, NJ 08231
(609) 652-1665

Great Swamp
Pleasant Plains Road
RD 1, Box 152
Basking Ridge, NJ 07920
(201) 647-1222

New Mexico

Bitter Lake
Box 7
Roswell, NM 88201
(505) 622-6755

Bosque del Apache
Box 1246
Socorro, NM 87801
(505) 835-1828

Includes Sevilleta unit
Las Vegas
Route 1, Box 399
Las Vegas, NM 87701
(505) 425-3581

Maxwell
Box 276
Maxwell, NM 87728
(505) 375-2331

San Andres
Box 756
Las Cruces, NM 88004
(505) 382-5047

New York

Iroquois
Box 517
Alabama, NY 14003
(716) 948-9154

Long Island NWR Complex
Box 21
Shirley, NY 11967
(516) 286-0485
Includes Amagansett,
Conscience Point, Elizabeth A.
Morton, Lido Beach, Oyster
Bay, Seatuck, Target Rock,
Wertheim units

Montezuma
3395 Route 5/20 East
Seneca Falls, NY 13148
(315) 568-5987

North Carolina

Alligator River
Box 1969
Manteo, NC 27954
(919) 473-1131
Includes Pea Island unit

Mackay Island (North
Carolina and Virginia)
Box 31
Knotts Island, NC 27950
(919) 429-3100
Includes Currituck unit

Mattamuskeet
Route 1, Box N-2
Swanquarter, NC 27885
(919) 926-4021
Includes Cedar Island, Pungo,
Swanquarter units

Pee Dee
Box 780
Wadesboro, NC 28170
(704) 694-4424

United States National Wildlife Refuges

Pocosin Lakes
Route 1, Box 195B
Creswell, NC 27928

Roanoake River
102 Dundee Street
Windsor, NC 27983
(919) 794-5326

North Dakota

Arrowwood Complex
Rural Route 1
Pingree, ND 58476
(701) 285-3341

Audubon Complex
Rural Route 1
Coleharbor, ND 58531
(701) 442-5474

Des Lacs Complex
Box 578
Kenmare, ND 58746
(701) 385-4046

Devils Lake Wetland
Management District
Box 908
Devils Lake, ND 58301
(701) 662-8611
Includes Lake Alice unit

J. Clark Salyer Complex
Box 66
Upham, ND 58789
(701) 768-2548

Kulm Wetland
Management District
Box E
Kulm, ND 58456
(701) 647-2866

Long Lake Complex
Rural Route Box 23
Moffit, ND 58560
(701) 387-4397

Tewaukon Complex
Rural Route 1, Box 75
Cayuga, ND 58013
(701) 724-3598

Upper Souris
Rural Route 1
Foxholm, ND 58738
(701) 468-5467

Ohio

Ottawa
14000 West State Route 2
Oak Harbor, OH 43449
(419) 898-0014

Oklahoma

Little River
General Delivery
Broken Bow, OK 74962
(405) 584-6211

Salt Plains
Route 1, Box 76
Jet, OK 73749
(405) 626-4794

Sequoyah
Route 1, Box 18A
Vian, OK 74962
(918) 773-5251

Tishomingo
Route 1, Box 151
Tishomingo, OK 73460
(405) 371-2402

WHERE TO BIRD

Washita
Route 1, Box 68
Butler, OK 73625
(405) 664-2205
Includes Optima unit

Wichita Mountains
Route 1, Box 448
Indiahoma, OK 73552
(405) 429-3221

Oregon

Malheur
Box 245
Princeton, OR 97721
(503) 493-2612

Oregon Coastal Refuges
2030 Marine Science Drive
Newport, OR 97365
(503) 867-0207

Sheldon/Hart Mountain
Complex
Post Office Building
Box 111, Room 308
Lakeview, OR 97630
(503) 947-3315

Umatilla (Oregon and
Washington)
Box 239
Umatilla, OR 97882
(503) 922-3232
Includes these units: Cold
Springs, McKay Creek

Western Oregon Refuge
Complex
26208 Finley Refuge Road
Corvallis, OR 97333
(503) 757-7236
Includes these units: Ankeny,
Bandon Marsh, Baskett Slough,
Cape Meares, William L. Finley

Pennsylvania

Erie
RD 1, Wood Duck Lane
Guy Mills, PA 16327
(814) 789-3585

Tinicum National
Environmental Center
Scott Plaza 2, Suite 104
Philadelphia, PA 19113
(215) 521-0662

Puerto Rico

Caribbean Islands
Box 510, Carr. 301, KM 5.4
Boqueron, PR 00622
(809) 851-7258
Includes these units: Buck
Island (Virgin Islands), Cabo
Rojo (Puerto Rico), Culebra
(Puerto Rico), Desecheo
(Puerto Rico), Green Cay
(Virgin Islands), Sandy Point
(Virgin Islands)

Rhode Island

Ninigret
Shoreline Plaza
Route 1A, Box 307
Charlestown, RI 02813
(401) 364-9124
Includes these units: Block
Island, Pettaquamscutt Cove,
Sachuest Point, Salt Meadow,
Trustom Pond

United States National Wildlife Refuges

South Carolina

Ace Basin
Box 840
Yemassee, SC 29945
(803) 846 9110

Cape Romain
390 Bulls Island Road
Awendaw, SC 29429
(803) 928-3368

Carolina Sandhills
Route 2, Box 330
McBee, SC 29101
(803) 335-8401

Santee
Route 2, Box 66
Summerton, SC 29148
(803) 478-2217

South Dakota

Huron Wildlife Management
District
Federal Building, Room 113
200 Fourth Street, SW
Huron, SD 57358
(605) 352-7014

Lacreek
HWC 3, Box 14
Martin, SD 57551
(605) 685-6508

Lake Andes
Rural Route 1, Box 77
Lake Andes, SD 57356
(605) 487-7683
Includes Karl E. Mundt unit

Madison Wetland
Management District
Box 48
Madison, SD 57042
(605) 256-2974

Sand Lake
Rural Route 1, Box 25
Columbia, SD 57433
(605) 885-6320

Waubay
Rural Route 1, Box 79
Waubay, SD 57273
(605) 947-4521

Tennessee

Cross Creeks
Route 1, Box 229
Dover, TN 37058
(615) 232-7477

Hatchie
Box 187
Brownsville, TN 38012
(901) 772-0501
Includes these units:
Chickasaw, Lower Hatchie,
Sunk Lake

Reelfoot
Route 2, Highway 157
Union City, TN 38261
(901) 538-2481
Includes Lake Isom unit

Tennessee
Box 849
Paris, TN 38242
(901) 642-2091

Texas

Anahuac
Box 278
Anahuac, TX 77514
(409) 267-3337

Includes these units: McFaddin, Moody, Texas Point
Aransas/Matagorda
Box 100
Austwell, TX 77950
(512) 286-3559

Attwater Prairie Chicken
Box 518
Eagle Lake, TX 77434
(409) 234-5940

Balcones Canyonlands
300 East Eighth Street, Room 873
Austin, TX 78701
(512) 482-5700

Brazoria
Box 1088
Angleton, TX 77515
(409) 849-6062
Includes these units: Big Boggy, San Bernard

Buffalo Lake
Box 228
Umbarger, TX 79091
(806) 499-3382

Hagerman
Route 3, Box 123
Sherman, TX 75090
(214) 786-2826

Laguna Atascosa
Box 450
Rio Hondo, TX 78583
(512) 748-3607

Lower Rio Grande Valley/Santa Ana Complex
320 North Main Street
McAllen, TX 78501
(512) 630-4636

Muleshoe
Box 549
Muleshoe, TX 79347
(806) 946-3341

Utah

Bear River Migratory Bird Refuge
Box 459
Brigham City, UT 84302
(801) 723-5887

Fish Springs
Box 568
Dugway, UT 84022
(801) 522-5353

Ouray
1680 West Highway 40, Room 1220
Vernal, UT 84078
(801) 723-5887

Vermont

Missisquoi
Route 2
Swanton, VT 05488
(802) 868-4781

United States National Wildlife Refuges

Virginia

Back Bay/Plum Tree Island
Box 6286
4005 Sandpiper Road
Virginia Beach, VA 23456
(804) 721-2412

Chincoteague/Wallops Island
Box 62
Chincoteague, VA 23336
(804) 336-6122

Eastern Shore of Virginia
RFD 1, Box 122B
Cape Charles, VA 23310
(804) 331-2760

Great Dismal Swamp (Virginia and North Carolina)
3100 Desert Road
Box 349
Suffolk, VA 23434
(804) 986-3705

Mason Neck
14416 Jefferson Davis
Highway, Suite 20A
Woodbridge, VA 22191
(703) 690-1297

Presquile/James River
Box 620
Hopewell, VA 23860
(804) 458-7541

Washington

Columbia
Drawer F
44 South Eighth Avenue
Othello, WA 99344
(509) 488-2668

Nisqually/San Juan Islands
100 Brown Farm Road
Olympia, WA 98506
(206) 753-9467

Ridgefield
Box 457
301 North Third Street
Ridgefield, WA 98642
(206) 887-4106
Includes these units: Conboy Lake, Pierce, Toppenish, Turnbull

Umatilla (*See* Oregon)

Willapa/Lewis and Clark
Ilwaco, WA 98624
(206) 484-3482
Includes Columbian White-tailed Deer unit

West Virginia

Ohio River Islands
Federal Building, Room 2711
425 Juliana Street
Parkersburg, WV 26102
(304) 420-7568

Wisconsin

Horicon
West 4279 Headquarters Road
Mayville, WI 53050
(414) 387-2658

Necedah
Star Route West, Box 386
Necedah, WI 54646
(608) 565-2551

Trempealeau
Route 1
Trempealeau, WI 54661
(608) 539-2311

Upper Mississippi River
La Crosse District
Box 415
La Crosse, WI 54601
(608) 784-3910

Wyoming

National Elk Refuge
Box C
Jackson, WY 83001
(307) 733-9212

Seedskadee
Box 67
Green River, WY 82935
(307) 875-2187

BUREAU OF LAND MANAGEMENT SITES

The Bureau of Land Management (BLM) was organized in 1946 as a division of the Department of the Interior. Its mission is to administer the extensive public lands of the United States. Since most BLM areas are vast tracts located in the western part of the country, the agency maintains only twelve state offices and about 60 district offices. In addition, the 337 million acres under the BLM's jurisdiction are divided into 170 resource area offices.

To get information about birding on BLM lands, start by contacting the state office. You may be referred on to a district office or a resource area office. The office staff can provide you with information about the area in which you are interested, including maps, brochures, and details about camping sites, birding hotspots, and recreational activities. BLM sites of varying sizes are scattered throughout the West—always request maps. Many BLM sites are leased out to private industry for ranching or for extractive activities such as logging and mining; access to these sites may be restricted or require prior permission. Some BLM sites are so inhospitable, dangerous, or ecologically fragile that access may be limited or require a permit. Always check first and be sure to comply with any requirements.

Bureau of Land Management Sites

National Office
Bureau of Land Management
U.S. Department of the Interior
1849 C Street, NW
Washington, DC 20240
(202) 208-5717

Regional Office (all states bordering and east of the Mississippi River)
BLM Eastern States Office
350 South Pickett Street
Alexandria, VA 22304
(703) 274-1369

State Offices
Alaska
BLM State Office
222 West Seventh Avenue
Anchorage, AK 99513
(907) 271-5555

Arizona
BLM State Office
Box 16563
3707 North Seventh Street
Phoenix, AZ 85011

California
BLM State Office
Federal Building
2008 East Cottage Way
Sacramento, CA 95825
(916) 978-4746

Colorado
BLM State Office
2850 Youngfield Street
Lakewood, CO 80125
(303) 236-1700

Idaho
BLM State Office
3380 Americana Terrace
Boise, ID 83706
(208) 334-1771

Mississippi
BLM State Office
300 Woodrow Wilson Drive
Jackson, MS 39213
(601) 965-4405

Montana (includes North Dakota and South Dakota)
BLM State Office
Box 36800
222 North 32nd Street
Billings, MT 59107
(406) 657-6561

Nevada
BLM State Office
Box 12000
Federal Building
850 Harvard Way
Reno, NV 89520
(702) 784-5311

New Mexico (includes Kansas, Oklahoma, and Texas)
BLM State Office
Box 1449
Federal Building
Santa Fe, NM 87504
(505) 988-6316

Oregon (includes Washington)
BLM State Office
Box 2965
825 Northeast Multnomah Street

Portland, OR 97208
(503) 231-6277

Utah
BLM State Office
324 South State Street
Salt Lake City, UT 84111
(801) 524-3146

Wyoming (includes Nebraska)
BLM State Office
Box 1828
2515 Warren Avenue
Cheyenne, WY 82003
(307) 772-2111

CANADIAN NATIONAL PARKS

Banff National Park, the first unit of Canada's extensive national park system, was founded in 1885. Today there are 34 national parks scattered among every province and both territories. The system includes such birding paradises as Point Pelee in Ontario (a famous migration hotspot), Wood Buffalo National Park in Alberta and Northwest Territories (Canada's largest park and the only nesting site of the whooping crane), and many coastal parks—such as Forillon in Quebec—famous for seabirds and bald eagles. Most of Canada's national parks offer superb hiking and canoeing, and most have campgrounds available on a first-come, first-served basis. Disabled persons have access to many facilities, including some trails. Modest fees for motor vehicle entry and such specialized activities as golfing are charged. To get up-to-date information about opening dates, best times to visit, facilities, and park activities, contact the parks or the regional offices.

For general information about Canada's national parks, contact:

Inquiry Centre
Environment Canada
Hull, Quebec K1A 0H3
(819) 997-2800

For more specific information about parks within a particular region, contact both the regional office and the park office.

Canadian National Parks

Western Region

Parks Service Western
Region Office
Room 520
220 Fourth Avenue South East
Calgary, Alberta T2P 3H8
(403) 292-4440

British Columbia

Glacier National Park
Box 350
Revelstoke, British Columbia
V0E 2S0
(604) 837-7500

Gwaii Haanas (South
Moresby) National Park
Reserve
Box 37
Queen Charlotte City, British
Columbia V0T 1S0
(604) 559-8818

Kootenay National Park
Box 220
Radium Hot Springs, British
Columbia V0A 1M0
(604) 347-9615

Mount Revelstoke
National Park
Box 350
Revelstoke, British Columbia
V0E 2S0
(604) 837-7500

Pacific Rim National Park
Box 280
Ucluelet, British Columbia
V0R 3A0
(604) 726-7721

Yoho National Park
Box 99
Field, British Columbia
V0A 1G0
(604) 343-6324

Alberta

Banff National Park
Box 900
Banff, Alberta T0L 0C0
(403) 762-1500

Elk Island National Park
Site 4, RR 1
Fort Saskatchewan, Alberta
T8L 2N7
(403) 992-6380

Jasper National Park
Box 10
Jasper, Alberta T0E 1E0
(403) 852-6161

Waterton Lakes National Park
Waterton, Alberta T0K 2M0
(403) 859-2224

Wood Buffalo National Park
(also in Northwest Territories)
(403) 872-2349

Prairie And Northern Region

Parks Service Prairie and
Northern Region Office
457 Main Street
Winnipeg, Manitoba R3B 3E8
(204) 983-2110

WHERE TO BIRD

Northwest Territories

Aulavik National Park
Box 1840
Inuvik, Northwest Territories
X0E 0T0
(403) 979-3248

Auyuittuq National Park
Box 1720
Iqaluit, Northwest Territories
X0A 0H0
(819) 979-6277

Ellesmere Island National Park
Box 1720
Iqaluit, Northwest Territories
X0A 0H0
(819) 979-6277

Ivvavik (formerly Northern Yukon) National Park
Box 1840
Inuvik, Northwest Territories
X0E 0T0
(403) 979-3248

Nahanni National Park
Postal Bag 300
Fort Simpson, Northwest Territories X0E 0N0
(403) 695-3151

North Baffin National Park
Box 1720
Iqaluit, Northwest Territories
X0A 0H0
(819) 979-6277

Wood Buffalo National Park
(also in Alberta)
Box 750
Fort Smith, Northwest Territories X0E 0P0
(403) 872-2349

Yukon Territory

Kluane National Park
Box 5495
Haines Junction, Yukon
Y0B 1L0
(403) 634-2251

Saskatchewan

Grasslands National Park
Box 150
Val Marie, Saskatchewan
S0N 2T0
(306) 298-2257

Prince Albert National Park
Box 100
Waskesiu Lake, Saskatchewan
S0J 2V0
(306) 663-5322

Manitoba

Riding Mountain
National Park
Wasagaming, Manitoba
R0J 2H0
(204) 848-2811

Ontario Region

Parks Service Ontario
Region Office
111 Water Street East
Cornwall, Ontario K6H 6S3
(613) 938-5866

Canadian National Parks

Ontario

Bruce Peninsula National Park
Bruce District
Box 189
Tobermory, Ontario N0H 2R0
(519) 596-2233

Georgian Bay Islands
National Park
Box 28
Honey Harbour, Ontario
P0E 1E0
(705) 756-2415

Point Pelee National Park
RR 1
Leamington, Ontario N8H 3V4
(519) 322-2371

Pukaskwa National Park
Highway 627, Hattie Cove
Heron Bay, Ontario P0T 1R0
(807) 229-0801

St. Lawrence Islands
National Park
RR 3, Two County Road Five
Mallory Town, Ontario
K0E 1R0
(613) 923-5261

Quebec Region

Parks Service Quebec
Region Office
3 Buade Street
Haute Ville
Quebec, Quebec G1R 4V7
(418) 648-4177

Quebec

Forillon National Park
Box 1220
122 Gaste Boulevard
Gaste, Quebec G0C 1R0
(418) 368-5505

La Mauricie National Park
Box 75A, Place Cascades
794 Fifth Street
Shawenegan, Quebec G9N 6V9
(819) 536-2638

Mingan Archipelago
National Park
Box 1180
1303 de la Digue
Havre-Saint-Pierre, Quebec
G0G 1P0
(418) 538-3331

Atlantic Region

Parks Service Atlantic
Region Office
Historic Properties
Upper Water Street
Halifax, Nova Scotia B3J 1S9
(902) 426-3457

New Brunswick

Fundy National Park
Alma, New Brunswick
E0A 1B0
(506) 887-2000

Kouchibouguac National Park
Kent County, New Brunswick
E0A 2A0
(506) 876-2443

Prince Edward Island

Prince Edward Island
National Park
Two Palmers Lane
Charlottetown, Prince Edward
Island C1A 5V6
(902) 566-7050

Nova Scotia

Cape Breton Highlands
National Park
Ingonish Beach, Nova Scotia
B0C 1L0
(902) 285-2270

Kejimkujik National Park
Box 236
Maitland Bridge
Annapolis County, Nova
Scotia B0T 1B0
(902) 242-2772

Newfoundland

Gros Morne National Park
Box 130
Rocky Harbour,
Newfoundland A0K 4N0
(709) 458-2417

Terra Nova National Park
Glovertown, Newfoundland
A0G 2L0
(709) 533-2801

Canadian Wildlife Service Sites

The Canadian Wildlife Service (CWS) is responsible for wildlife habitat and nonconsumptive wildlife recreation. Since birding tops the list of nonconsumptive wildlife recreation, the staff at the provincial CWS offices can be very helpful. The CWS oversees 45 National Wildlife Areas (NWA), 99 Migratory Bird Sanctuaries (MBS), and 30 Ramsar Sites (RS). National Wildlife Areas are usually on federally owned lands. Access to these areas may be restricted; check with the regional office before visiting. Access to most Migratory Bird Sanctuaries is by permit only; some MBSs are on private property. When requesting a visitor's permit, plan at least two months in advance. Ramsar Sites are designated wetlands areas of particular importance as waterfowl habitat. (The name comes from the Iranian town where the international agreement was reached in 1971.) Almost all these sites are located above 60° N. Ramsar Sites can be very large—the Whooping Crane Summer Range site in Alberta/Northwest Territories covers some 6.7 million acres. Some are within national parks, NWAs, and MBSs; others are separate areas, often on private land. Because most Ramsar Sites are located in fragile tundra habitat, permits are almost always necessary.

In addition to the national park system, the provinces and territories of Canada contain many outstanding parks, refuges, sanctuaries, and other good sites for birding. Contact the provincial or territorial wildlife offices for details, addressing your query to the nongame biologist. Staff members in these offices often have detailed local knowledge and can be very helpful to birders.

The Canadian Wildlife Service is a division of Environment Canada, the agency in overall charge of environmental affairs. For more information about visiting sites in Canada, contact the regional offices of CWS/Environment Canada and the provincial wildlife agencies.

ENVIRONMENT CANADA INFORMATION CENTRES

Environment Canada
Quebec Region
351 St. Joseph Boulevard
Hull, Quebec K1A 0H3
(819) 997-2800

Environment Canada
Atlantic Region
45 Alderney Drive
Dartmouth, Nova Scotia
B2Y 2N6
(902) 426-7990

Environment Canada
Ontario Region
25 St. Clair Avenue East
Toronto, Ontario M4T 1M2
(416) 973-6467

Environment Canada,
Western and Northern Region
Twin Atria 2, 2nd Floor
4999 98th Avenue
Edmonton, Alberta T6B 2X3
(403) 468-8074
or
1901 Victoria Avenue,
Room 241
Regina, Saskatchewan S4P 3R4
(306) 780-6002

Environment Canada Pacific
and Yukon Region
Kapilano 100, 3rd Floor
Park Royal South
West Vancouver, British
Columbia V7T 1A2
(604) 666-5900

CANADIAN WILDLIFE SERVICE REGIONAL OFFICES

Pacific and Yukon Region

Canadian Wildlife Service
5421 Robertson Road
Delta, British Columbia
V4K 3Y3
(604) 946-8546

Western and Northern Region

Canadian Wildlife Service
Twin Atria 2, 2nd Floor
4999 98th Avenue
Edmonton, Alberta T6B 2X3

Ontario

Canadian Wildlife Service
1725 Woodward Drive
Ottawa, Ontario K1A 0H3
(613) 998-4693

Quebec

Canadian Wildlife Service
1141 Route de l'Eglise
Ste. Foy, Quebec G1V 4H5
(418) 648-3914

Atlantic Region

Canadian Wildlife Service
Box 1590
Sackville, New Brunswick
E0A 3C0
(506) 536-3025

CWS NATURAL AREAS

National Wildlife Areas

Alberta

Blue Quilts
Meanook
Spiers Lake

British Columbia

Alasken
Columbia
Qualicum
Vaseux-Bighorn
Widgeon Valley

Manitoba

Pope
Rockwood

New Brunswick

Cape Jourimain
Portage Island
Portobello Creek
Shepody
Tintamarre

Nova Scotia

Boot Island
Chiqnecto
Port Joli/Port Hebert
Sand Pond
Sea Wolf Island
Wallace Bay

Ontario

Big Creek
Eleanor Island
Long Point
Mississippi Lake
Mohawk Island
Prince Edward Point
Scotch Bonnet Island
St. Clair
Wellers Bay
Wye Marsh

Quebec

Cap Tourmente
Îles de Contrecoeur
Îles de l'Estuaire
Îles de la Paix
La Baie de l'Isle-Verte
Lac Saint-François
Pointe de l'Est
Pointe-au-Père

Saskatchewan

Bradwell
Last Mountain Lake
Prairie
Raven Island
Stalwart
St. Denis
Tway
Webb

Northwest Territories

Polar Bear Pass

Migratory Bird Sanctuaries

Alberta

Inglewood
Red Deer
Richardson Lake
Saskatoon Lake

British Columbia

Christie Islet
Esquimalt Lagoon
George C. Reifel
Nechako River
Shoal Harbour
Vaseux Lake
Victoria Harbour

New Brunswick

Grand Manan
Machias Seal Island

Newfoundland and Labrador

Terra Nova

Nova Scotia

Amherst Point
Big Grace Bay
Haley Lake
Kentville
Port Joli/Port Herbert
Sable Island
Sable River

Ontario

Beckett Creek
Chantry Island

WHERE TO BIRD

Eleanor Island
Fielding
Guelph
Hanna Bay
Mississippi Lake
Moose River
Pinafore Park
Rideau
Southern James Bay
St. Joseph's Island
Upper Canada
Young Lake

Prince Edward Island

Black Pond

Quebec

Baie des Loups
Betchouane
Bird Rocks
Boatswain Bay
Bonaventure Island/Percé Rock
Brador Bay
Brandy Pot
Cap Saint-Ignace
Carillon Island
Corossol Island
Couvee Islands
Île Blanche
Île á la Brume
Île aux Basques
Île aux Fraises
Île aux Herons
Îles Pelerins
Îles de la Paix
Îles Sainte-Marie
Île Saint-Ours
Kamouraska Island
L'Islet
L'Isle-Verte
Montmagny
Mont Saint-Hilaire
Nicolet
Philipsburg
Saint-Augustin
Saint-Omer
Saint-Vallier
Senneville
Toris-Saumons
Watshishou

Saskatchewan

Basin and Middle Lake
Duncairn Reservoir
Indian Head
Last Mountain Lake
Lenore Lake
Murray Lake
Neely Lake
Old Wives Lake
Opuntia Lake
Redberry Lake
Scent Grass Lake
Sutherland
Upper Rousay Lake
Val Marie Reservoir
Wascana Lake

Northwest Territories

Akimiski Island
Anderson River Delta
Banks Island One
Banks Island Two
Bylot Island
Cape Dorset
Cape Parry
Dewey Soper
East Bay

Harry Gibbons
Kendall Island
McConnell River
Queen Maud Gulf
Seymour Island

Ramsar Sites

Alberta

Beaverhill Lake
Hay-Zama Lakes
Peace-Athabaska Delta
Whooping Crane
Summer Range

British Columbia

Alasken

Manitoba

Delta Marsh
Oak-Hammock Marsh

New Brunswick

Shepody Bay/Mary's Point

Newfoundland and Labrador

Grand Codroy Estuary

Nova Scotia

Chiqnecto
Musquodoboit Harbour
Outlet Estuary
Southern Bight/Minas Basin

Ontario

Long Point
Point Pelee National Park
Polar Bear Provinicial Park
Southern James Bay
St. Clair

Prince Edward Island

Malpeque Bay

Quebec

Cap Tourmente
La Baie de l'Isle-Verte
Lac François

Saskatchewan

Last Mountain Lake
Quill Lakes

Northwest Territories

Dewey Soper
Mcconnell Rover
Polar Bear Pass
Queen Maud Gulf
Rasmussen Lowlands

Yukon

Old Crow Flats

PROVINCIAL AGENCIES

British Columbia

Wildlife Management
Ministry of Environment
780 Blanshard Street
Victoria, British Columbia
V8V 1X5
(604) 387-9728

Alberta

Fish and Wildlife Division
Department of Forestry, Lands
and Wildlife
Petroleum Plaza, North Tower
9945 108th Street
Edmonton, Alberta T5K 2G6
(403) 427-5192

Saskatchewan

Wildlife Branch
Saskatchewan Parks,
Renewable Resources
3211 Albert Street
Regina, Saskatchewan S4S 5W6
(306) 787-2314

Manitoba

Wildlife Branch
Department of
Natural Resources
1495 St. James Street
Winnipeg, Manitoba R3H 0W9
(204) 945-6809

Ontario

Wildlife Branch
Ministry of Natural Resources
99 Wellesley Street West
Whitney Block, Queen's Park,
Room 4620
Toronto, Ontario M7A 1W3
(416) 965-4252

Quebec

Wildlife Branch
Department of Recreation,
Hunting and Fishing
150 St. Cyrille Boulevard East
Quebec, Quebec G1R 4Y1
(418) 644-8377

New Brunswick

Fish and Wildlife Branch
Department of Natural
Resources and Energy
Box 6000
Fredericton, New Brunswick
E3B 5H1
(506) 453-2433

Nova Scotia

Wildlife Division
Department of Lands
and Forest
Box 516
Kentville, Nova Scotia
B4N 3X3
(902) 678-8921

Prince Edward Island

Fish and Wildlife Division
Department of the
Environment
Box 2000
Charlottetown, Prince Edward
Island C1A 7N8
(902) 368-4683

Newfoundland

Wildlife Division
Department of Environment
and Lands
Box 8700
St. John's, Newfoundland
A1B 4J6
(709) 576-2817

Canadian Provincial Agencies

Yukon Territory

Wildlife Branch
Department of Renewable
Resources
Yukon Territorial Government
Box 2703
Whitehorse, Yukon Y1A 2C6
(403) 667-5786

Northwest Territories

Director of Wildlife
Management
Department of Renewable
Resources
Government of the Northwest
Territories
Box 1320
Yellowknife, Northwest
Territories X1A 2L9
(403) 873-7411

THE NATURE CONSERVANCY PRESERVES

Since its founding in 1951, The Nature Conservancy has been a leading force in preserving natural habitats and biodiversity. A private, nonprofit organization, The Nature Conservancy takes a pragmatic approach to conservation, working with individuals, government agencies at all levels, corporations, foundations, universities, and other conservation organizations. Today, The Nature Conservancy protects millions of acres in the United States and Canada; it is helping to protect millions more in the Caribbean, Latin America, and elsewhere in the hemisphere. With more than 1,300 preserves, The Nature Conservancy is the largest private system of nature sanctuaries in the world.

The indescribably valuable work of The Nature Conservancy is an amazing boon to birders. Because The Nature Conservancy is especially interested in protecting sites that are biologically rich and contain endangered species, its preserves are excellent places for birding. More than 9,000 birders a year visit the Cape May Migratory Bird Refuge in New Jersey, for example.

Any trip to a Nature Conservancy site must be planned in advance. Many sites are fairly easy to get to and are open to the public on a routine basis. Many other sites, however, are closed to the public, sometimes because they are easements on private property, but usually because the site is ecologically fragile. Other sites are in very rugged areas and are inaccessible all or some of the year, while others can be reached only by boat. Some sites are open on a limited or seasonal basis. The well-known Ramsey Canyon site in Arizona, for example, requires advance reservations and is closed part of the week.

A voluntary donation is usually requested of visitors to Nature Conservancy preserves. Give generously.

The best approach to visiting The Nature Conservancy sites is to write directly to the relevant state field office. The staff will respond with detailed information about the preserves as well as information about field trips and other activities. In the listings below, the names, addresses, and phone numbers of some individual preserves that are of particular interest to birders are also given where available.

International Headquarters

The Nature Conservancy
1815 North Lynn Street
Arlington, VA 22209
(703) 841-5300

Regional Offices

The Nature Conservancy
California Regional Office
785 Market Street
San Francisco, CA 94103
(415) 777-0487

The Nature Conservancy
Eastern Regional Office
201 Devonshire Street, 5th floor
Boston, MA 02110
(617) 542-1908

The Nature Conservancy
Florida Regional Office
2699 Lee Road, Suite 500
Winter Park, FL 32789
(407) 628-5887

The Nature Conservancy
Latin America Division
1815 North Lynn Street
Arlington, VA 22209
(703) 841-5300

The Nature Conservancy
Midwest Regional Office
1313 Fifth Street SE, Suite 314
Minneapolis, MN 55414
(612) 331-0700

The Nature Conservancy
New York Regional Office
1736 Western Avenue
Albany, NY 12203
(518) 869-6959

The Nature Conservancy
Pacific Regional Office

The Nature Conservancy Preserves

1116 Smith Street, Suite 201
Honolulu, HI 96817
(808) 537-4508

The Nature Conservancy
Southeast Regional Office
Box 2267
Chapel Hill, NC 27515
(919) 967-5493

The Nature Conservancy
Western Regional Office
2060 Broadway, Suite 230
Boulder, CO 80302
(303) 444-1060

State Field Offices

Alabama

The Nature Conservancy
Alabama Field Office
2821-C Second Avenue South
Birmingham, AL 35233
(205) 251-1155

Alaska

The Nature Conservancy
Alaska Field Office
601 West Fifth Avenue,
Suite 550
Anchorage, AK 99501
(907) 276-3133
Note: No preserves

Arizona

The Nature Conservancy
Arizona Field Office
300 East University Boulevard,
Suite 230
Tucson, AZ 85705
(602) 622-3861

Preserves:

Aravaipa Canyon Preserve
Klondyke Station
Willcox, AZ 85643
(602) 828-3443

Hassayampa River
Box 1162
Wickenburg, AZ 85358
(602) 684-2772

Mile High/Ramsey Canyon
RR 1, Box 84
Hereford, AZ 85615
(602) 378-2785

Muleshoe Ranch
RR 1, Box 1542
Willcox, AZ 85643
(602) 586-7072

Patagonia Sonoita Creek
Preserve
Box 815
Patagonia, AZ 85624
(602) 394-2400

Ramsey Canyon Preserve
27 Ramsey Canyon Road
Hereford, AZ 85615
(602) 378-2785

San Pedro River
(602) 622-3861

Arkansas

The Nature Conservancy
Arkansas Field Office
300 Spring Building, Suite 717
Little Rock, AR 72201
(501) 372-2750

Preserves:

Baker Prairie Nature Area

Cossatot River State Park and
Natural Area
Route 1, Box 170A
Wickes, AR 71973
(501) 385-2201

Lorance Creek Natural Area

Moro Creek Bottoms
Natural Area

Nacatoch Ravines
Natural Area

Stone Road Glade Natural Area

Terre Noir Blackland Prairie
Natural Area

Warren Prairie Natural Area

California

California Regional Office
785 Market Street
San Francisco, CA 94103
(415) 777-0487

Preserves:

Big Bear Valley Preserve
Box 1418
Sugarloaf, CA 92386
(714) 866-4190

Big Morongo Canyon Preserve
Box 780
Morongo Valley, CA 92256
(619) 363-7190

Carrizo Plain Natural Area
Box 3098
California Valley, CA 93453
(805) 475-2360

Coachella Valley Preserve
Box 188
Thousand Palms, CA 92276
(619) 343-1234

Cosumnes River Preserve
6500 Desmond Road
Galt, CA 95632
(916) 684-2816

Elkhorn Slough Reserve
Elkhorn Slough Foundation
Box 267
Moss Landing, CA 95039
(408) 728-2822

Fairfield Osborn Preserve
6543 Lichau Road
Penngrove, CA 94951
(707) 795-5069

Gray Davis Dye Creek Preserve
11010 Foothill Road
Los Molinos, CA 96055

Jepson Prairie Preserve
Kern River Preserve
Box 1662
Weldon, CA 93283
(619) 378-2531

The Nature Conservancy Preserves

Lanphere-Christensen
Dunes/Mad River Slough
Preserve
6800 Lanphere Road
Arcata, CA 95521
(707) 822-6378

Landels-Hill Big Creek Preserve
McCloud River Preserve
Box 409
McCloud, CA 96057

Nipomo Dunes Preserve
Box 15810
San Luis Obispo, CA 93406
(805) 545-9925

Northern California Coast
Range Preserve
42101 Wilderness Road
Branscomb, CA 95417
(707) 984-6653

Ring Mountain Preserve
3152 Paradise Drive
Tiburon, CA 94920
(415) 435-6465

Sacramento River Oxbow
Preserve
c/o Dr. Syd Thomas
Route 2, Box 188
Chico, CA 95926
(916) 343-3185
(916) 891-7380

Santa Cruz Island Preserve
213 Stearns Wharf
Santa Barbara, CA 93101
(805) 962-9111

Santa Rosa Plateau Preserve
22115 Tenaja Road
Murrieta, CA 92562
(714) 677-6951

Colorado

The Nature Conservancy
Colorado Program
1244 Pine Street
Boulder, CO 80302
(303) 444-2950

Preserve:

Phantom Canyon Preserve

Connecticut

The Nature Conservancy
Connecticut Field Office
55 High Street
Middletown, CT 06457
(203) 344-0716

Selected preserves:

Helen G. Altschul Preserve

Harry E. Barnes Memorial
Preserve and Nature Center

Bellamy Preserve

Chapman Pond Area

Dennis Farm Preserve

Devil's Den Preserve
Box 1162
Weston, CT 06883
(203) 226-4991

Iron Mountain Reservation

Lord Cove Preserve

Poquetanuck Cove Preserve

Rock Spring Wildlife Refuge

Selden Creek Preserve

Sunny Valley Preserve
6 Sunny Valley Lane
New Milford, CT 06776
(203) 355-3716

Turtle Creek Wildlife Sanctuary

Weir Preserve
Box 7033
Wilton, CT 06897

Delaware

The Nature Conservancy
Delaware Field Office
319 South State Street
Dover, DE 19903
(302) 674-3550

Preserve:

Port Mahon

Florida

The Nature Conservancy
Florida Regional Office
2699 Lee Road, Suite 500
Winter Park, FL 32789
(407) 628-5887

Preserves:

Blowing Rocks Preserve
Box 3795
Tequesta, FL 33469
(407) 575-2297

Florida Keys
(305) 296-3880

Matanzas Pass Preserve
Box 2692
Fort Myers, FL 33932

Theodore Roosevelt Preserve

Tiger Creek Preserve
Central Florida Land Steward
225 East Stuart Avenue
Lake Wales, FL 33853
(813) 676-0521

Georgia

The Nature Conservancy
Georgia Field Office
4725 Peachtree Street NE,
Suite 136
Atlanta, GA 30309
(404) 873-6946

Preserves:

Heggies Rock Preserve

Marshall Forest Preserve

Great Basin Region

The Nature Conservancy
Great Basin Field Office
Box 11486, Pioneer Station
Salt Lake City, UT 84147
(801) 531-0999

Hawaii

The Nature Conservancy
Hawaii Field Office
1116 Smith Street, Suite 201
Honolulu, HI 96817
(808) 537-4508

The Nature Conservancy Preserves

Preserves:
Kamakou Preserve
Box 220
Kualapuu, HI 96757
(808) 553-5236

Waikamoi Preserve
Box 1716
Makawao, HI 96768
(808) 572-7849

Kapunakea Preserve
Box 1716
Makawao, HI 96768
(808) 572-7849

Idaho

The Nature Conservancy
Idaho Field Office
Box 165
Sun Valley, ID 83353
(208) 726-3007

Preserves:

Big Wood River

Birds of Prey Preserve

Chilly Slough

Formation Springs Preserve

Gamble Lake Preserve

Garden Creek Preserve

Snake River Route
Lewiston, ID 83501
(509) 243-4055

Hixson Sharptail Preserve

Idler's Rest Preserve

Silver Creek Preserve
Box 624
Picabo, ID 83348
(208) 788-2203

South Fork
of the Snake Preserve

Stapp-Soldier Creek Preserve

Thousand Springs Preserve
1205 Thousand Springs Grade
Wendell, ID 83355
(208) 536-6797

Illinois

The Nature Conservancy
Illinois Field Office
79 West Monroe Street,
Suite 900
Chicago, IL 60603
(312) 346-8166

Preserves:

Cache River Wetlands

Cedar Glen Eagle
Roost Preserve
(217) 256-4519

Indian Boundary Prairies

Nachusa Grassland Preserve

North Branch Prairie Project

Indiana

The Nature Conservancy
Indiana Field Office
1330 West 38th Street
Indianapolis, IN 46208
(317) 923-7547

Preserves:

Buffalo Flat

Chapman Lake Wetlands

Conley Wildlife Refuge

Elkhart Bog

WHERE TO BIRD

Fern Cliff
Grand Kankakee Marsh
(219) 552-9614

Hemlock Bluff

Little Calumet-Langeluttig Marsh

Mallard Roost

Manitou Island Wetlands

Maumee Woods

North Branch Elkhart Wetlands

Pipewort Pond Nature Preserve

Portland Arch

Shoup-Parsons Swamp Woods

Thornhill Farm
(219) 344-1225
Tribbett Woods

Iowa

The Nature Conservancy
Iowa Field Office
431 East Locust, Suite 200
Des Moines, IA 50309
(515) 244-5044

Preserves:

Ames High Prairie Preserve

Behrens Pond and Woodland Preserve

Berry Woods Preserve

Brayton-Horsley Prairie Preserve

Cedar Hills Sand Prairie Preserve

Crossman Prairie Preserve

The Diggings Preserve

Five Ridge Preserve

Freda Haffner Preserve

Greiner Family Preserve

Hanging Box Preserve

Hoffman Prairie Preserve

Lock and Dam #14 Eagle Area

Retz Memorial Forest Preserve

Savage Memorial Woods Preserve

Silver Lake Fen Preserve

Sioux City Prairie Preserve

Steele Prairie Preserve

Williams Prairie Preserve

Kansas

The Nature Conservancy
Kansas Field Office
Southwest Plaza Building
3601 Southwest 29th Street
Topeka, KS 66614
(913) 272-5115

Preserve:

Konza Prairie

Kentucky

The Nature Conservancy
Kentucky Field Office
642 West Main Street
Lexington, KY 40508
(606) 259-9655

Preserves:

Bad Branch Preserve

Boone County Cliffs Preserve

The Nature Conservancy Preserves

Brigadoon Preserve

Lilley Cornett Woods Preserve

Mantle Rock Preserve

Metropolis Lake Preserve

Pilot Knob Preserve

Louisiana

The Nature Conservancy
Louisiana Field Office
Box 4125
Baton Rouge, LA 70812
(504) 338-1040

Preserves:

Charter Oak Nature Preserve

Lake Cocodrie Preserve

Schoolhouse Springs Nature Preserve

Tunica Hills Preserve

White Kitchens Nature Preserve

Maine

The Nature Conservancy
Maine Field Office
14 Main Street, suite 401
Brunswick, ME 04011
(207) 729-5181

Preserves:

Great Wass Island Preserve

Indian Point Blagden
Indian Point
Mount Desert, ME 04660
(207) 288-4838

Waterboro Barrens
(207) 729-5181

Maryland

The Nature Conservancy
Maryland Field Office
Chevy Chase Metro Building
Two Wisconsin Circle,
Suite 600
Chevy Chase, MD 20815
(301) 656-8673

Preserves:

Battle Creek Cypress
Swamp Sanctuary
Courthouse
Prince Frederick, MD 20678
(301) 535-5327

Choptank Wetlands

Cranesville Swamp

Finzel Swamp

Helen Creek Hemlock Preserve

Nassawango Creek
Furnace Town Foundation
Box 207
Snow Hill, MD 21863
(301) 632-2032

Otwell Woodland

Robinson Neck/Frank M. Ewing Preserve

Selinger Marsh

Third Haven Woods

Massachusetts

The Nature Conservancy
Massachusetts Field Office
201 Devonshire Street, 5th floor
Boston, MA 02110
(617) 423-2545

WHERE TO BIRD

Preserve:

Black Pond Preserve
Massachusetts Audubon
Southeast Regional Center
2000 Main Street
Marshfield, MA 02050
(617) 837-9400

Michigan

The Nature Conservancy
Michigan Field Office
2840 East Grand River, suite 5
East Lansing, MI 48823
(517) 332-1741

Preserves:

Colonial Point Forest Preserve
Dickinson Island Preserve
Erie Marsh Preserve
Grand Mere State Park
Grass River Natural Area
Harbor Island Preserve
Nordhouse Dunes Preserve
Skegemog Swamp Wildlife Area
Walkinshaw Wetlands Preserve

Minnesota

The Nature Conservancy
Minnesota Field Office
1313 Fifth Street, SE
Minneapolis, MN 55414
(612) 331-0750

Preserves:

Helen Allison Savanna Preserve
Black Dog Fen Preserve
Blue Devil Valley Preserve
Bluestem Prairie Preserve
Paul Bunyan Savanna Preserve
Chippewa Prairie Preserve
Clinton Prairie Preserve
Cold Springs Preserve
Egret Island Preserve
Felton Prairie Complex
Foxhome and Kettledrummer Prairies Preserve
Hole-in-the-Mountain Prairie Preserve
Kasota Prairie Preserve
Langley River Preserve
MacDougall Homestead Preserve
North Heron Lake Preserve
Norway Dunes Preserve
Ordway Prairie Preserve
Pankratz Memorial Prairie Preserve
Pembina Trail Preserve
Plover Prairie Preserve
Red Rock Prairie Preserve
Regal Meadow Preserve
Roscoe Prairie Preserve
Schaefer Prairie Preserve
Seven Sisters Prairie Preserve
Staffanson Prairie Preserve
Strandness Prairie Preserve
Susie Island Preserve
Wabu Woods Preserve
Weaver Dunes Preserve
Western Prairie North Preserve
Zimmerman Prairie Preserve

The Nature Conservancy Preserves

Mississippi

The Nature Conservancy
Mississippi Field Office
Box 1028
Jackson, MS 39215
(601) 355-5357

Preserves:

Black Creek Swamp
Clark Creek Natural Area
(601) 362-9612

Dahomey National
Wildlife Refuge

Hillside National
Wildlife Refuge
(601) 965-4465

Pascagoula River Wildlife
Management Area
Box 451
Jackson, MS 39205
(601) 947-6376

Shipland Wildlife
Management Area

Missouri

The Nature Conservancy
Missouri Field Office
2800 South Brentwood
Boulevard
St. Louis, MO 63144
(314) 968-1105

Selected preserves:

Bennett Spring
Savanna Preserve

Cook Meadow Preserve

Dobbins Woodland Preserve

Hunkah Prairie Preserve

Lichen Glade Preserve

Marmaton River Bottoms Wet
Prairie Preserve

Jamerson C. McCormack
Loess Mounds Preserve

Mo-Ko Prairie Preserve

Monegaw Prairie Preserve

Pawhuska Prairie Preserve

Shut-In Mountain Fen Preserve

Victor Glade Preserve

Wah-Kon-Tah Prairie Preserve

Wah-Sha-She Prairie Preserve

Zahorsky Woods Preserve

Montana

The Nature Conservancy
Montana Field Office
32 South Ewing
Helena, MT 59601
(406) 443-0303

Preserve:

Pine Butte Preserve
Pine Butte Guest Ranch
HC 58, Box 34C
Choteau, MT 59422
(406) 466-2377

Nebraska

The Nature Conservancy
Nebraska Field Office
1722 St. Mary's Avenue,
Suite 403
Omaha, NE 68102
(402) 342-0282

Preserves:

Niobrara Valley Preserve
Route 1, Box 348
Johnstown, NE 69214
(402) 722-4440

Willa Cather Memorial Prairie

New Hampshire

The Nature Conservancy
New Hampshire Field Office
2-1/2 Beacon Street, suite 6
Concord, NH 03301
(603) 224-5853

Preserves:

Frank Bolles Nature Reserve

Hurlbert Swamp

Stamp Act Island Preserve

Barry Lougee, Preserve Steward
Hodge Shore Road
Wolfeboro, NH 03894
(603) 569-3446

Warwick Preserve

West Branch Pine Barrens Preserve

New Jersey

The Nature Conservancy
New Jersey Field Office
Elizabeth D. Kay
Environmental Center
200 Pottersville Road
Chester, NJ 07930
(908) 879-7262

Preserves:

Bennett Bogs Preserve

Cape May Migratory Bird Refuge

New Mexico

The Nature Conservancy
New Mexico Field Office
212 East Marcy Street
Santa Fe, NM 87501
(505) 988-3867

Preserves:

Dripping Springs Natural Area
(505) 522-1219

Gila Riparian Preserve

Rattlesnake Springs Preserve

New York

The Nature Conservancy
New York Regional Office
1736 Western Avenue
Albany, NY 12203
(518) 869-6959

New York Chapter Offices

Adirondack Nature Conservancy
Box 65
Keene Valley, NY 12943
(518) 576-2082

Clintonville Pine Barrens

Coon Mountain

Everton Falls Preserve

Gadway Sandstone Pavement Barrens

Silver Lake Camp Preserve

The Nature Conservancy Preserves

Central and Western New York Chapters

315 Alexander Street, suite 301
Rochester, NY 14604
(716) 546-8030

Baltimore Woods Preserve
Center for Nature Education
Box 133
Marcellus, NY 13108
(315) 673-1350

Deer Lick Preserve
Derby Hill Bird Observatory
Onondaga Audubon Society
Sage Creek Road
Mexico, NY 13114
(315) 963-8291

Eldridge Wilderness Preserve

Lake Julia Preserve

Moss Lake Preserve

Noyes Preserve
117 North Way
Camillus, NY 13031
(315) 963-8291

Eastern New York Chapter
1736 Western Avenue
Albany, NY 12203
(518) 869-0453

Bear Swamp

Christman Sanctuary

Stewart Preserve

Hannacroix Ravine

Kenrose Sanctuary

Limestone Rise

Long Island Chapter
250 Lawrence Hill Road
Cold Spring Harbor, NY 11724
(516) 367-3225

Butler-Huntington Woods

Daniel Davis Sanctuary

East Farm Preserve

Husing Farm Preserve

Hope Goddard Iselin Preserve

Kempf/Reppa Preserve

Thorne Preserve

Uplands Farm Sanctuary

David Weld Sanctuary
Lower Hudson Chapter
223 Katonah Avenue
Katonah, NY 10536
(914) 232-9431

A.W. Butler Memorial Sanctuary

Thompson Pond Preserve
South Fork/
Shelter Island Chapter
Box 2694
Sag Harbor, NY 11963
(516) 725-2936

Preserves:

Accabonac Harbor Preserve

Mashomack Preserve
Box 410
Shelter Island, NY 11964
(516) 749-1001

Barcelona Neck

Scallop Pond Preserve

Wolf Swamp Preserve

North Carolina

The Nature Conservancy
North Carolina Field Office
Carr Mill Mall, suite 223
Carrboro, NC 27510
(919) 967-7007

Preserves:

Bat Cave Preserve

Big Yellow Mountain Preserve

Bluff Mountain Preserve

Camassia Slopes Preserve

Carolina Bays Preserve

Green Swamp Preserve

Lanier Quarry
Savanna Preserve

Lower Roanoake
River Preserve

Nags Head Woods Preserve

North Dakota

The Nature Conservancy
North Dakota Chapter Office
1014 East Central Avenue
Bismarck, ND 58501
(701) 222-8464

Preserve:

Cross Ranch Nature Preserve
HC 2, Box 150
Hensler, ND 58547
(701) 794-8741

Ohio

The Nature Conservancy
Ohio Field Office
1504 West First Avenue
Columbus, OH 43212
(614) 486-4194

Preserves:

Big Darby Creek
(614) 486-4194

Browne Lake Bog Preserve

Buzzardroost Rock Preserve

J. Arthur Herrick Fen Preserve

Lynx Prairie Preserve

Oklahoma

The Nature Conservancy
Oklahoma Field Office
320 South Boston, suite 1222
Tulsa, OK 74103
(918) 585-1117

Preserves:

Arkansas River
Least Tern Preserve

Bald Eagle
Winter Roosting Preserve

Canadian River
Least Tern Preserve

Redbud Valley Nature Preserve

Tallgrass Prairie Preserve
(918) 287-4803

The Nature Conservancy Preserves

Oregon

The Nature Conservancy
Oregon Field Office
1205 NW 25th Avenue
Portland, OR 97210
(503) 228-9561

Preserve:

Cascade Head Natural Area

Pennsylvania

The Nature Conservancy
Pennsylvania Field Office
1211 Chestnut Street, 12th floor
Philadelphia, PA 19107
(215) 963-1400

Preserves:

Bristol Marsh Preserve

Goat Hill Serpentine Barrens Preserve

Long Pond Preserve

Tannersville Cranberry Bog Preserve

Monroe County Conservation District
RD 2, Box 2335A
Stroudsburg, PA 18360
(717) 992-7334

Thompson Wetlands
c/o Patricia Christian
Box 24
Starlight, PA 18461
(717) 278-1174

Woodbourne Forest and Wildlife Sanctuary
RD 6, Box 6294
Montrose, PA 18801
(717) 278-3384

Rhode Island

The Nature Conservancy
Rhode Island Field Office
45 South Angell Street
Providence, RI 02906
(401) 331-7110

Preserves:

Block Island
Box 1287
Block Island, RI 02807
9401) 466-2129

Goosewing Beach Preserve
Norman Bird Sanctuary
(401) 846-2577

South Carolina

The Nature Conservancy
South Carolina Field Office
Box 5475
Columbia, SC 29250
(803) 254-9049

Preserves:

Flat Creek Nature Area and Forty-Acre Rock Preserve

Peachtree Rock Preserve

Washo Reserve

South Dakota

The Nature Conservancy
South Dakota Chapter Office
196 East Sixth Street
Sioux Falls, SD 57117
(701) 222-8464

Selected preserves:

Clovis Prairie Preserve

Crystal Springs Centennial Prairie Preserve

Hansen Nature Preserve
Samuel H. Ordway Memorial Prairie Preserve
HCR 1, Box 16
Leola, SD 57456
(605) 439-3475

Sioux Prairie Preserve
Wilson Savanna Preserve

Tennessee

The Nature Conservancy
Tennessee Field Office
2002 Richard Jones Road, Suite 304C
Nashville, TN 37215
(615) 298-3111
Note: The nine preserves in Tennessee are not open to the public.

Texas

The Nature Conservancy
Texas Field Office
Box 1440
San Antonio, TX 78295
(512) 224-8774

Preserves:

Clymer Meadow Preserve

Roy Larsen Sandylands Sanctuary
Box 909
Silsbee, TX 77656
(409) 385-4135

Lennox Woods Preserve
Texas Hill Country
Tridens Prairie

Utah

The Nature Conservancy
Great Basin Field Office
Box 11486, Pioneer Station
Salt Lake City, UT 84102
(801) 531-0999

Preserves:

Cunningham Ranch

Deep Creek Basin

Graham Ranch Preserve

Layton Marsh

Lytle Ranch Preserve
Box 398
Santa Clara, UT 84765
(801) 378-2289

Matheson Wetlands Preserve
Ruby Valley Preserves
Stillwater Wetlands

Vermont

The Nature Conservancy
Vermont Field Office
27 State Street
Montpelier, VT 05602
(802) 229-4425

Preserves:
Barr Hill Nature Preserve
LaPlatte River Marsh Preserve
Newark Pond Preserve
Shaw Mountain Natural Area

Virginia

The Nature Conservancy
Virginia Field Office
1233A Cedars Court
Charlottesville, VA 22903
(804) 295-6106

Preserves:
Alexander Berger
Memorial Sanctuary
200A Fauquier Street
Fredericksburg, VA 22401
(703) 371-8324

Falls Ridge Preserve
Route 2, Box 289
Christianburg, VA 24703
(703) 382-2220

Fraser Preserve
2126 North Rolfe Street
Arlington, VA 22209
(703) 528-4952

Helena's Island Preserve
Route 1, Box 242
Shipman, VA 22971
(703) 862-0056

Virginia Coast Reserve
Brownsville
Nassawadox, VA 23413
(804) 442-3049

Washington

The Nature Conservancy
Washington Field Office
217 Pine Street, Suite 1100
Seattle, WA 98101
(206) 343-4344

Preserves:
Willapa Bay Bioreserve
(206) 343-4344
Yellow Island Preserve

West Virginia

The Nature Conservancy
West Virginia Field Office
723 Kanawha Boulevard East,
Suite 500
Charleston, WV 25301
(304) 345-4350

Selected preserves:
Brush Creek Preserve
Cranesville Swamp Preserve

Wisconsin

The Nature Conservancy
Wisconsin Field Office
333 West Mifflin, Suite 107
Madison, WI 53703
(608) 251-8140

Preserves:

Baxter's Hollow Preserve

Mink River Estuary Preserve

Pine Hollow Preserve

Thomson and Thousand's
Rock Point Prairies Preserve

Wyoming

The Nature Conservancy
Wyoming Field Office
258 Main Street, suite 200
Lander, WY 82520
(307) 332-2971

Sweetwater River Project

Tensleep Preserve

NATIONAL AUDUBON SOCIETY SITES

A conservation organization with more than 550,000 members, the National Audubon Society works toward the preservation and wise use of America's natural heritage. The society's roots go back to 1886, when George Bird Grinnell, editor of the magazine *Forest and Stream*, formed the first American bird association, the Audubon Society, to protest the wholesale slaughter of birds by market gunners and for their plumes. The public was so responsive that Grinnell was overwhelmed, and had to disband the association almost as soon as it was started. Others took up the challenge, however, and formed state societies. The Massachusetts and Pennsylvania Audubon Societies were formed in 1896; by 1899, fifteen other states also had organizations.

These separate groups formed the National Committee of Audubon Societies, an informal alliance, in 1901. In 1905, they formally created the National Association of Audubon Societies for the Protection of Wild Birds and Animals. The name was changed to National Audubon Society in 1940.

From the beginning, the society has been involved with education and conservation. In 1900 the society began hiring wardens to patrol and protect nesting and breeding sites, such as Mantinicus Rock in Maine. Three wardens were killed in the line of duty. The first national wildlife refuge was created in 1903 on Pelican Island in Florida at the urging of the society. The Audubon refuge system began around then too. The largest Audubon refuge, the 26,000-acre Paul J. Rainey Sanctuary in Louisiana, was acquired in 1924.

For more information about National Audubon Society activities, and for information about state Audubon societies, see Chapter 4.

AUDUBON SANCTUARIES

The National Audubon Society sanctuary system protects more than 250,000 acres; in addition, thousands of acres are protected by local chapters. The list below is of just some of the National Audubon Society sites usually open to the public. Some Audubon sites are closed to protect habitat, and others are closed simply because they are inaccessible all or part of the year. Many of the sanctuaries open to the public require prior permission and advance arrangements, and many sanctuaries ask for modest entrance fees. It is always best to contact the site well ahead of a planned visit. For more sanctuary information contact:

National Audubon Society
Sanctuary Department
93 West Cornwall Road
Sharon, CT 06069
(203) 364-0048

Alabama

Dauphin Island Sanctuary
Box 189
Dauphin Island, AL 365228
(205) 861-2882

Arizona

Appleton-Whittell Research Ranch Sanctuary
Box 44
Elgin, AZ 85611
(602) 455-5522

California

Bobelaine
Sacramento Audubon Society
3615 Auburn Boulevard
Sacramento, CA 95821

Richardson Bay Wildlife Sanctuary and Whittell Education Center
376 Greenwood Beach Road
Tiburon, CA 94920
(415) 388-2524

WHERE TO BIRD

Starr Ranch Audubon
Sanctuary
Box 967
Trabuco Canyon, CA 92678
(714) 858-0309

Paul L. Wattis Audubon
Sanctuary
National Audubon Society
555 Audubon Place
Sacramento, CA 95825
(916) 481-5332

Williams Sisters Ranch
Sanctuary
California Sanctuaries,
National Audubon Society
555 Audubon Place
Sacramento, CA 95825
(916) 481-5332

Connecticut

Audubon Center in Greenwich
and Fairchild Wildflower
Garden
613 Riversville Road
Greenwich, CT 06831
(203) 869-5272

Guilford Salt Meadows
Sanctuary
330 Mulberry Point Road
Guilford, CT 06437
(203) 458-9981

Miles Wildlife Sanctuary
95 West Cornwall Road
Sharon, CT 06069
(203) 364-5302

Northeast Audubon Center
RR 1, Box 171
Sharon, CT 06069
(203) 364-0520

Florida

Corkscrew Swamp Sanctuary
Route 6, Box 1875A
Sanctuary Road
Naples, FL 33964
(813) 657-3771

Cowpens Key
115 Indian Mound Trail
Tavernier, FL 33070
(305) 852-5092

Kitchen Creek Wildlife
Sanctuary
Sanctuary Chairman
Box 6762
West Palm Beach, FL 33405

Lake Okeechobee
c/o Corkscrew Swamp
Sanctuary
Route 6, Box 1875A
Sanctuary Road
Naples, FL 33964
(813) 657-3771

Lake Worth Island
c/o Kitchen Creek Sanctuary
Box 6762
West Palm Beach, FL 33405

Ordway-Whittell Kissimmee
Prairie Sanctuary
c/o Corkscrew Swamp
Sanctuary
Route 6, Box 1875A

Sanctuary Road
Naples, FL 33964
(813) 467-8497

Rookery Bay Sanctuary
c/o Corkscrew Swamp
Sanctuary
3967 North Road
Naples, FL 33942
(813) 774-2922

Tampa Bay
410 Ware Boulevard, Suite 500
Tampa, FL 33619
(813) 623-6826

Kentucky

Clyde E. Buckley Wildlife
Sanctuary
1305 Germany Road
Frankfort, KY 40601
(606) 873-5711

Jefferson County Memorial
Forest
c/o Metro Parks Office
1297 Trevilian Way
Louisville, KY 40201
(502) 459-0440

Louisiana

Rainey Wildlife Sanctuary
Route 5, Box 1990
Abbeville, LA 70510
(318) 893-4703

Maine

Borestone Mountain Audubon
Wildlife Sanctuary
Box 112

Monson, ME 04464
(207) 997-3607 (summer)
(207) 997-3558 (winter)

Allan D. Cruickshank
Wildlife Sanctuary
(607) 529-5828 (summer)
(607) 257-7308 (winter)

Duryea Morton Wildlife
Sanctuary
c/o Todd Wildlife Sanctuary
Keene Neck Road
Medomak, ME 04551
(207) 529-5148

Little Duck Island
(607) 529-5828 (summer)
(607) 257-7308 (winter)

Matinicus Rock
(607) 529-5828 (summer)
(607) 257-7308 (winter)

Edgar B. Mulford Wildlife
Sanctuary
c/o Todd Wildlife Sanctuary
Keene Neck Road
Medomak, ME 04551
(207) 529-5148

P.W. Sprague
Memorial Sanctuary
Box 3163
Prout's Neck, ME 04074
(607) 529-5828 (summer)
(607) 257-7308 (winter)

Ten Pound Island
(607) 529-5828 (summer)
(607) 257-7308 (winter)

Todd Wildlife Sanctuary
Keene Neck Road
Medomak, ME 04551
(207) 529-5148

Western Egg Rock
c/o Todd Wildlife Sanctuary
Keene Neck Road
Medomak, ME 04551
(207) 529-5148

Maryland

Nanjemoy Marsh Sanctuary
Southern Maryland
Audubon Society
Box 181
Bryans Road, MD 20616
(301) 375-8552

Minnesota

Audubon Center of the
Northwoods
Route 1
Sandstone, MN 55072
(612) 245-2648

Nebraska

Niobara River Sanctuary
c/o West Central
Regional Office
200 South Wind Place,
Suite 205
Manhattan, KS 66502
(913) 537-4385

Lillian Annette Rowe
Sanctuary
Route 2, Box 112A
Gibbon, NE 68840
(308) 468-5282

New Mexico

Randall Davey
Audubon Center
Box 9314
Santa Fe, NM 87504
(505) 983-4609

New York

Buttercup Wildlife Sanctuary
c/o Miles Wildlife Sanctuary
95 West Cornwall Road
Sharon, CT 06069
(203) 364-5302

Constitution Marsh Sanctuary
RFD 2, Route 9D
Garrison, NY 10524
(914) 265-2601

Ruth Walgreen Franklin and
Winifred Fels Audubon
Sanctuaries
Bedford Audubon Society
Box 322
Mt. Kisco, NY 10549
(914) 666-6177

Livingston Marsh Sanctuary
Northern Catskills
Audubon Society
Box 395A
Bethel Ridge Road
Catskill, NY 12414

Theodore Roosevelt Memorial
Bird Sanctuary
134 Cove Road
Oyster Bay, NY 11771
(516) 922-3200

Audubon Sanctuaries

Scully Sanctuary
306 South Bay Avenue
Islip, NY 11751
(516) 227-4289

North Carolina

North Carolina Coastal
Island Sanctuary
Box 5223
Wrightsville Beach, NC 28480
(919) 256-3779

Pine Island Sanctuary
Box 174
Poplar Branch, NC 27965
(919) 453-2838

North Dakota

Edward M. Brigham III Alkali
Lake Sanctuary
RR 1, Box 79A
Spiritwood, ND 58481
(701) 252-3822

Ohio

Aullwood Audubon Center
and Farm
1000 Aullwood Road
Dayton, OH 45414
(513) 890-7360

Oregon

Ten Mile Creek Sanctuary
Audubon Western
Regional Office
555 Audubon Place
Sacramento, CA 95825
(916) 481-5332

Pennsylvania

Crosswicks Wildlife Sanctuary
Wyncote Audubon Society
161 Greenwood Avenue
Wyncote, PA 19095

South Carolina

Francis Beidler Forest
Route 1, Box 600
Harleyville, SC 29448
(803) 462-2150

Silver Bluff Plantation
Sanctuary
Route 1, Box 391
Jackson, SC 29831
(803) 827-0781

Texas

Robert Porter Allen Sanctuary
(512) 541-8034
Green Island
c/o Robert Porter Allen
Sanctuary
(512) 541-8034

North Deer Island
Fort Crockett, Building 311
Galveston, TX 77550
(409) 744-2665

Sabal Palm Grove
Box 5052
Brownsville, TX 78523
(512) 541-8034

South Bird Island
c/o Robert Porter Allen
Sanctuary
(512) 541-8034

Sydnes Island
c/o Robert Porter Allen
Sanctuary
(512) 541-8034

Vingt-et-un Islands
Star Route 2, Box 734
Anahuac, TX 77514
(409) 355-2252

West Bay Bird Island
Fort Crockett, Building 311
Galveston, TX 77550
(409) 744-2665

Wisconsin

Hunt Hill Sanctuary
Friends of Hunt Hill
Audubon Sanctuary
RR 1, Box 285
Sarona, WI 54870

Schlitz Audubon Center
1111 East Brown Deer Road
Milwaukee, WI 53217
(414) 352-2880

Chapter 2

TRAVEL INFORMATION FOR BIRDERS

Know before you go is the guiding principle for planning a birding trip, even a local excursion. Of course, knowing where the birds are is of crucial importance, but equally important is knowledge about the locale itself—the availability of campgrounds, lodgings, food, and other amenities for the traveling birder. Fortunately, such information is available in detailed abundance—and it's very often free.

STATE WILDLIFE OFFICES

Every state has an agency that deals with fish, game, and wildlife. The name of the agency varies from state to state, but it usually includes the words conservation, game, or wildlife somewhere in the title. These agencies are staffed by dedicated and enthusiastic professionals who are often fonts of detailed local knowledge about birding sites in their state. You'll get more from your local birding if you contact the biologists at your own state's wildlife office. (Some states have ornithologists on staff, but generally the person to contact is the nongame biologist.) If you're planning an out-of-state trip, be sure to get in touch with the target state's wildlife office. The staff will probably be able to provide maps, brochures, checklists, and other information about state and local parks, refuges, sanctuaries, and other birding sites—all free or very inexpensive. They may also have information about state and local research projects, birding clubs, and events and programs of interest to visiting birders. And of course, if you are interested in fishing, hunting, or any other outdoor activities, they'll have information about all that as well.

Some large states have regional wildlife offices. The addresses and phone numbers given here are for the main office, which is almost always located in the state capital.

Alabama

Department of Conservation
and Natural Resources
Game and Fish Division
64 North Union Street
Montgomery, AL 36104
(205) 242-3465

Alaska

Alaska Department of Wildlife
Conservation
1225 West 8th Street
Box 3-2000
Juneau, AK 99802
(907) 465-4190

Arizona

Arizona Game and Fish
Commission
2222 West Greenway Road
Phoenix, AZ 85023
(602) 942-3000

Arkansas

Arkansas Game and Fish
Commission
Information Department
Two Natural Resources Drive
Little Rock, AR 72205
(501) 223-6351

California

California Department of Fish
and Game
1416 Ninth Street
Box 944209
Sacramento, CA 94244
(916) 445-3531

Colorado

Colorado Division of Wildlife
6060 Broadway
Denver, CO 80216
(303) 297-1802

Connecticut

Department of
Environmental Protection
Wildlife Bureau Headquarters
State Office Building
165 Capitol Avenue
Hartford, CT 06106
(203) 566-4683

Delaware

Department of Natural
Resources and
Environmental Control
Division of Fish and Wildlife
Richardson and
Robbins Building

State Wildlife Offices

89 Kings Highway
Box 1401
Dover, DE 19903
(302) 736-4580

Florida

Department of
Natural Resources
Game and Freshwater Fish
Commission
Farris Bryant Building
Tallahassee, FL 32301
(904) 488-1960

Georgia

Department of
Natural Resources
Game and Fish Division
2258 Northlake Parkway
Tucker, GA 30084
(404) 656-3510

Hawaii

Department of Land and
Natural Resources
Division of Forestry and
Wildlife
Kalanimoku Building
1151 Punchbowl Street
Honolulu, HI 96813
(808) 548-2861

Idaho

Idaho Department of Fish
and Game
600 South Walnut
Box 25
Boise, ID 83707
(208) 334-3700

Illinois

Division of Wildlife Resources
Lincoln Tower Plaza
524 South Second Street
Springfield, IL 62706
(217) 782-6384

Indiana

Department of
Natural Resources
Division of Fish and Wildlife
607 State Office Building
Indianapolis, IN 46204
(317) 232-4080

Iowa

Iowa Conservation
Commission
Wallace State Office Building
Des Moines, IA 50319
(515) 281-8174

Kansas

Kansas Department of
Wildlife and Parks
Headquarters Office
Route 2, Box 54A
Pratt, KS 67124
(316) 672-5911

Kentucky

Kentucky Department of
Fish and Wildlife Resources
One Game Farm Road
Frankfort, KY 40601
(502) 564-4336

TRAVEL INFORMATION FOR BIRDERS

Louisiana

Department of Wildlife
and Fisheries
Box 15570
Baton Rouge, LA 71360
(504) 765-2934

Maine

Maine Department of Inland
Fisheries and Wildlife
284 State Street
Augusta, ME 04333
(207) 287-2871

Maryland

Department of Natural
Resources
Forest, Park and Wildlife
Service
Tawes State Office Building
Annapolis, MD 21401
(301) 269-3195

Massachusetts

Division of Fisheries
and Wildlife
Leverett Saltonstall Building
Government Center
100 Cambridge Street
Boston, MA 02202
(617) 727-3151

Michigan

Michigan Department of
Natural Resources
Wildlife Division
Box 30028
Lansing, MI 48909
(517) 373-1263

Minnesota

Minnesota Department of
Natural Resources
Section of Wildlife
500 Lafayette Road
St. Paul, MN 55155
(612) 296-3344

Mississippi

Department of
Wildlife Conservation
Box 451
Jackson, MS 39205
(601) 961-5300

Missouri

Missouri Department of
Conservation
Box 180
Jefferson City, MO 65102
(314) 751-4115

Montana

Department of Fish,
Wildlife and Parks
1420 East Sixth Street
Helena, MT 59620
(406) 444-2535

Nebraska

Nebraska Game and
Parks Commission
2200 North 33rd Street
Box 30370
Lincoln, NE 68503
(402) 464-0641

State Wildlife Offices

Nevada

Nevada Department of Wildlife
Box 10678
Reno, NV 89520
(702) 789-0500

New Hampshire

New Hampshire Fish
and Game
Two Hazen Drive
Concord, NH 03301
(603) 271-3211

New Jersey

New Jersey Division of Fish,
Game and Wildlife
CN 400
Trenton, NJ 08625
(609) 292-2965

New Mexico

New Mexico Department of
Game and Fish
Villagra Building
State Capitol
Santa Fe, NM 87503
(505) 827-7911

New York

New York State Department of
Environmental Conservation
Bureau of Wildlife
50 Wolf Road
Albany, NY 12233
(518) 474-2121

North Carolina

North Carolina Wildlife
Resources Commission
Archdale Building
512 North Salisbury Street
Raleigh, NC 27611
(208) 334-3700

North Dakota

North Dakota Game and
Fish Department
100 North Bismarck
Expressway
Bismarck, ND 58501
(701) 221-6300

Ohio

Ohio Division of Wildlife
Fountain Square, Building C-4
Columbus, OH 43224
(614) 265-6305

Oklahoma

Oklahoma Department of
Wildlife Conservation
Game Division,
Nongame Program
1801 North Lincoln
Box 53465
Oklahoma City, OK 73152
(405) 521-3855

Oregon

Oregon Department of
Fish and Wildlife
506 Southwest Mill Street
Box 3503
Portland, OR 97208
(503) 229-5403

TRAVEL INFORMATION FOR BIRDERS

Pennsylvania

Pennsylvania Game
Commission
8000 Derry Street
Box 1567
Harrisburg, PA 17120
(717) 787-3633

Rhode Island

Department of Environmental
Management
Division of Fish and Game
Government Center
Tower Hill Road
Wakefield, RI 02879
(401) 789-3094

South Carolina

South Carolina Wildlife and
Marine Resources Department
Division of Wildlife and
Freshwater Fisheries
Robert C. Dennis Building
Box 167
Columbia, SC 29202
(803) 758-0001

South Dakota

Game, Fish and Parks
Headquarters
445 East Capitol
Pierre, SD 57501
(605) 773-3485

Tennessee

Tennessee Wildlife Resource
Agency
Ellington Agricultural Center
Box 40747
Nashville, TN 37204
(615) 360-0500

Texas

Texas Parks and Wildlife
Department
4200 Smith School Road
Austin, TX 78744
(512) 479-4800

Utah

Utah Division of Wildlife
Resources
1596 West North Temple Street
Salt Lake City, UT 84116
(801) 533-9333

Vermont

Agency of Natural Resources
Fish and Wildlife Department
Montpelier, VT 05602
(802) 241-3670

Virginia

Department of Game and
Inland Fisheries
4101 West Broad Street
Box 11104
Richmond, VA 23230
(804) 257-1000

Washington

Washington Department
of Wildlife
600 North Capitol Way
Olympia, WA 98504
(206) 753-5700

Wisconsin

Department of Natural
Resources
101 South Webster Street
Madison, WI 53702
(608) 266-2277

Wyoming

Wyoming Game and
Fish Department
Information Section
Cheyenne, WY 82002
(307) 777-7735

TRAVEL AND TOURISM OFFICES

When planning a birding trip or vacation, contact the travel and tourism office (or the convention and visitors' bureau or chamber of commerce) for the country, state, province, region, or even city you will be visiting. These offices, many of which have convenient toll-free phone numbers, exist for the sole purpose of promoting travel to their particular region. The ever-helpful staff members will be delighted to send you piles of excellent, up-to-date general information about their area. Free maps and information about lodgings, campgrounds, and parks and other natural areas are readily available. You can also get information about such nonbirding attractions as recreational areas, theme parks, shopping malls, historic and cultural sites and events, restaurants, and the like—very important if nonbirders are along on the trip. As a general rule, much more information than you actually need is sent out as soon as it is requested.

UNITED STATES

Alabama

Alabama Bureau of
Tourism and Travel
401 Adams Avenue
Box 4309
Montgomery, AL 36103-4309
(800) ALABAMA
(205) 242-4169

Alaska

Alaska Division of Tourism
Box 110801
Juneau, AK 99811
(907) 465-2010

Alaska Marine
Highway System
Box R
Juneau, AK 99811
(907) 465-3941
(800) 642-0066

TRAVEL INFORMATION FOR BIRDERS

Fairbanks Visitor Information
Center
550 First Avenue
Fairbanks, AK 99701-4790
(907) 456-5774

Homer Chamber of Commerce
Homer, AK 99603
(907) 235-7740

Hyder Community Association
Box 149
Hyder, AK 99923
(907) 636-2708

Nome Convention
and Visitors Bureau
Box 251
Nome, AK 99762
(907) 443-5535

Sitka Convention and
Visitors Bureau
Box 1226
Sitka, AK 99835

St. Paul Island
(800) 544-2248

American Samoa

American Samoa Government
Office of Tourism
c/o McClellan Corporation
International
21318 Dumetz Road
Box 4070
Woodland Hills, CA 91365
(818) 884-0480

Arizona

Arizona Office of Tourism
1100 W. Washington Avenue
Phoenix, AZ 85007
(602) 542-8687

Bensun Chamber of Commerce
(602) 586-2841

Bisbee Chamber of Commerce
(602) 432-5421

Douglas Chamber of
Commerce
1125 Pan American
Douglas, AZ 85607
(602) 364-2477

Sierra Vista Chamber of
Commerce
77 Calle Portal
Sierra Vista, AZ 85635
(800) 288-3861
(602) 458-6940

Tombstone Chamber of
Commerce
(602) 457-9317

Wilcox Chamber of Commerce
(800) 200-2272

Arkansas

Arkansas Department of
Parks and Tourism
1 Capital Mall
Little Rock, AR 72201
(501) 371-7777
(800) NATURAL

Travel and Tourism Offices

California

California Office of Tourism
Box 9278
Van Nuys, CA 91409
(800) TO-CALIF

City of Huntington Beach
Public Information Office
2000 Main Street
Huntington Beach, CA 92648
(800) SAY-OCEAN
(714) 536-5511

Colorado

Colorado Tourism Board
Box 38700
Denver, CO 80238
(303) 592-5410
(800) COLORADO

Connecticut

Connecticut Department of
Economic Development
Tourism Division
865 Brook Street
Rocky Hill, CT 06067
(203) 258-4355
(800) CT-BOUND

Delaware

Delaware Tourism Office
99 Kings Highway
Box 1401
Dover, DE 19903
(302) 739-4271
(800) 441-8846

District of Columbia

Washington, DC Convention
and Visitors Association
1212 New York Avenue, NW
Washington, DC 20005
(202) 789-7000

Florida

Florida Division of Tourism
Visitor Inquiry Section
126 West Van Buren Street
Tallahassee, FL 32301
(904) 487-1462

Georgia

Georgia Department of
Industry, Trade, and Tourism
Box 1776
Atlanta, GA 30301-1776
(404) 656-3590
(800) VISIT-GA

Guam

Guam Visitors Bureau,
United States
c/o Richard Keating,
Marketing Representative
425 Madison Avenue
New York, NY 10017
(212) 888-4110
(800) 228-GUAM

Hawaii

Hawaii Visitors Bureau
Box 2359
Honolulu, HI 96804
(808) 586-2423

Idaho

Idaho Division of Tourism
Development
700 West State Street
Boise, ID 83720-2700
(208) 334-2470
(800) 635-7820

Illinois

Illinois Office of Tourism
310 South Michigan Avenue,
Suite 108
Chicago, IL 60604
(312) 793-2094
(800) 223-0121

Indiana

Indiana Division of Tourism
One North Capitol Avenue,
Suite 700
Indianapolis, IN 46204-2288
(317) 232-8860
(800) 289-6646

Iowa

Iowa Division of Tourism
200 East Grand Avenue
Des Moines, IA 50309
(515) 242-4705
(800) 345-IOWA

Kansas

Kansas Travel and
Tourism Division
400 Southwest Eighth Street,
5th Floor
Topeka, KS 66603-3957
(913) 296-2009
(800) 2-KANSAS

Kentucky

Kentucky Department of
Travel Development
Capitol Plaza Tower,
Suite 2200
500 Mero Street
Frankfort, KY 40601-1968
(502) 564-4930
(800) 225-TRIP

Louisiana

Louisiana Office of Tourism
Box 94291
Baton Rouge, LA 70804-9291
(504) 925-3860
(800) 33-GUMBO

Southwest Louisiana
Convention and
Visitors Bureau
1211 North Lakeshore Drive
Lake Charles, LA 70601
(800) 456-SWLA

Maine

Maine Office of Tourism
189 State Street
Augusta, ME 04333
(207) 289-2423
(800) 533-9595

Marianas Protectorate

Marianas Visitors Bureau
Box 861
Saipan, MP 96950
(670) 234-8325

Travel and Tourism Offices

Maryland

Maryland Office of Tourist
Development
217 East Redwood Street
Baltimore, MD 21202
(410) 333-6611
(800) 543-1036

Massachusetts

Massachusetts Office
of Travel and Tourism
100 Cambridge Street,
13th Floor
Boston, MA 02202
(617) 727-3201
(800) 447-MASS

Michigan

Michigan Travel Bureau
Department of Commerce
Box 30226
Lansing, MI 48909
(517) 373-1195
(800) 5432-YES

Sault Convention
and Visitors Bureau
2581 I-75 Business Spur
Sault Ste. Marie, MI 49783
(800) MI-SAULT

Minnesota

Minnesota Office of Tourism
375 Jackson Street
250 Skyway Level
Saint Paul, MN 55101-1848
(612) 296-5029
(800) 657-3700 (USA)
(800) 766-8687 (Canada)

Thief River Falls
Chamber of Commerce
2017 Highway 59 SE
Thief River Falls, MN 56701
(800) 827-1629

Winona Convention and
Visitors Bureau
Box 870
Winona, WI 55987
(800) 657-4972

Mississippi

Mississippi Division
of Tourism
Department of Economic
Development
Box 22825
Jackson, MS 39205
(601) 359-3297
(800) 647-2290

Missouri

Missouri Division of Tourism
Box 1055
Jefferson City, MO 65102
(314) 751-4133
(800) 877-1234

Montana

Travel Montana
Room 259
Deer Lodge, MT 59722
(406) 444-2654
(800) 541-1447

Nebraska

Nebraska Division of Travel
and Tourism
301 Centennial Mall South,
Room 88937
Lincoln, NE 68509
(402) 471-3796
(800) 228-4307

Nevada

Nevada Commission on
Tourism
Capitol Complex
Carson City, NV 89710
(702) 687-4322
(800) NEVADA-8

New Hampshire

New Hampshire Office of
Travel and Tourism
Box 856
Concord, NH 03302-0856
(603) 271-2343

New Jersey

New Jersey Division of
Travel and Tourism
Department of Commerce and
Economic Development
20 West State Street
CN 826
Trenton, NJ 08625
(609) 292-2470
(800) JERSEY-7

Cape May Chamber of
Commerce
Box 556
Cape May, NJ 08204
(609) 884-5508

New Mexico

New Mexico Department
of Tourism
1100 St. Francis Drive
Box 20003
Santa Fe, NM 87503
(505) 827-0291
(800) 545-2040

Socorro Chamber of
Commerce
Box 743
Socorro, NM 87801
(505) 835-0424
(505) 835-7600

New York

New York State Department of
Economic Development
Division of Tourism
One Commerce Plaza
Albany, NY 12245
(518) 474-4116
(800) CALL-NYS

North Carolina

North Carolina Division of
Travel and Tourism
430 North Salisbury Street
Raleigh, NC 27603
(919) 733-4171
(800) VISIT-NC

North Dakota

North Dakota Tourism
Promotion
Liberty Memorial Building
604 East Boulevard

Travel and Tourism Offices

Bismarck, ND 58505
(701) 224-2525
(800) HELLO-ND

Ohio

Ohio Division of Travel
and Tourism
Box 1001
Columbus, OH 43266-0101
(614) 466-8844
(800) BUCKEYE

Oklahoma

Oklahoma Tourism and
Recreation Department
Travel and Tourism Division
500 Will Rogers Building,
DA92
Oklahoma City, OK
73105-4492
(405) 521-3981
(800) 652-6552

Oregon

Oregon Economic
Development Department
Tourism Division
775 Summer Street, NE
Salem, OR 97310
(503) 373-1270
(800) 547-7842

Pennsylvania

Pennsylvania Bureau of Travel
Marketing
130 Commonwealth Drive
Warrendale, PA 15086
(717) 787-5453
(800) VISIT-PA

Puerto Rico

Puerto Rico Tourism
575 Fifth Avenue
New York, NY 10017
(800) 223-6530
(212) 599-6262

Puerto Rico Tourism Company
Box 5268
Miami, FL 33102
(800) 866-STAR, extension 17

Rhode Island

Rhode Island Tourism Division
Seven Jackson Walkway
Providence, RI 02903
(401) 277-2601
(800) 556-2484

South Carolina

South Carolina Division of
Tourism
Box 71, Room 902
Columbia, SC 29202-0071
(803) 734-0235

South Dakota

South Dakota Department of
Tourism
711 East Wells Avenue
Pierre, SD 57501-3369
(605) 773-3301
(800) 843-1930

Tennessee

Tennessee Department of
Tourism Development
Box 23170
Nashville, TN 37202
(615) 741-2158

Texas

Texas Department of Commerce
Tourism Division
Box 12728
Austin, TX 78711-2728
(512) 462-9191
(800) 88-88-TEX

Harlingen Area
Chamber of Commerce
311 East Tyler
Harlingen, TX 78550
(800) 531-7346

Kingsville Visitor Center
Box 1562
Kingsville, TX 78364
(800) 333-5032

Port Arthur Convention and Visitors Bureau
3401 Cultural Center Drive
Port Arthur, TX 77642
(800) 235-7822
(409) 985-7822

Rockport-Fulton
Chamber of Commerce
Rockport, TX 78382
(800) 242-0071

South Padre Island
Visitors Bureau
(800) 343-2368

United States Virgin Islands

U.S. Virgin Islands
Division of Tourism
Box 6400, VITIA
Charlotte Amalie, St. Thomas
USVI 00801
(809) 774-8784
(800) 372-8784

Virgin Islands
Division of Tourism
2655 Lejeune Road, Suite 907
Coral Gables, FL 33134
(305) 442-7200

Virgin Islands Division of Tourism
1270 Avenue of the Americas
New York, NY 10020
(212) 582-4520
(212) 581-3405

Utah

Utah Travel Council
Council Hall, Capitol Hill
Salt Lake City, UT 84114
(801) 538-1030
(800) UTAH-FUN

Vermont

Vermont Travel Division
134 State Street
Montpelier, VT 05602
(802) 828-3236
(800) 338-0189

Virginia

Virginia Tourism
1021 East Cary Street,
14th Floor
Richmond, VA 23219
(804) 786-4484
(800) VISIT-VA

Travel and Tourism Offices

Washington

Washington State Tourism
Development Division
Box 42513
Olympia, WA 98504-2513
(206) 586-2088
(800) 544-1800

West Virginia

West Virginia Division of
Tourism and Parks
2101 Washington Street East
Charleston, WV 25305
(304) 348-2286
(800) 225-5982

Wisconsin

Wisconsin Division of Tourism
Box 7606
Madison, WI 53707
(608) 266-2161
(800) 372-2737 (in state)
(800) 432-TRIP (out of state)

Wyoming

Wyoming Division of Tourism
I-25 at College Drive
Cheyenne, WY 82002
(307) 777-7777
(800) 225-5996

CANADA

Alberta

Alberta Tourism, Parks, and
Recreation
Vacation Counseling, 3rd
Floor, City Centre
10155 102 Street
Edmonton, Alberta T5J 4L6
(403) 427-4321
(800) 661-8888

British Columbia

Tourism British Columbia
Parliament Buildings
Victoria, British Columbia
V8V 1X4
(604) 685-0032
(800) 663-6000

Manitoba

Travel Manitoba
Department 20, 7th Floor
155 Carlton Street
Winnipeg, Manitoba R3C 3H8
(204) 945-3777
(800) 665-0040

New Brunswick

Tourism New Brunswick
Box 12345
Fredericton, New Brunswick
E3B 5C3
(506) 453-2444
(800) 561-0123

Newfoundland and Labrador

Newfoundland and Labrador
Department of Tourism
and Culture
Box 8730
Saint John's, Newfoundland
A1B 4K2
(709) 729-2830
(800) 563-6353

Northwest Territories

Northwest Territories Tourism
Box 1320
Yellowknife, Northwest
Territories, X1A 2L9
(403) 873-7200
(800) 661-0788

Nova Scotia

Department of Tourism
and Culture
Box 456
Halifax, Nova Scotia, B3J 2R5
(902) 424-5000
(800) 565-0000 (Canada)
(800) 341-6096 (USA)

Ontario

Ontario Travel
Queen's Park
Toronto, Ontario M7A 2R9
(416) 314-0944
(800) ONTARIO

Prince Edward Island

Tourism Prince Edward Island
Box 940
Charlottestown, Prince Edward
Island C1A 7M5
(902) 368-4444
(800) 565-0267

Quebec

Tourisme Quebec
C.P. 20 000
Quebec, Quebec G1K 7X2
(514) 873-2015
(800) 363-7777

Saskatchewan

Tourism Saskatchewan
1919 Saskatchewan Drive
Regina, Saskatchewan S4P 3V7
(306) 787-2300
(800) 667-7191

Yukon Territory

Tourism Yukon
Box 2703
Whitehorse, Yukon Y1A 2C6
(403) 667-5340

INTERNATIONAL

Anguilla

Anguilla Tourist
Information Office
271 Main Street
Northport, NY 11768
(800) 553-4939

Antigua and Barbuda

Antigua and Barbuda
Tourist Board
610 Fifth Avenue
New York, NY 10020
(212) 541-4117

Aruba

Aruba Tourism Authority
1000 Harbor Boulevard
Weehawken, NJ 07047
(201) 330-0800

Australia

Australia Tourist Commission
489 Fifth Avenue
New York, NY 10017
(800) 445-4400

Travel and Tourism Offices

Austria

Austrian National
Tourist Office
11601 Wilshire Boulevard,
Suite 2480
Los Angeles, CA 90025
(310) 477-3332

Austrian National
Tourist Office
500 Fifth Avenue, Suite 2009
New York, NY 10110
(212) 944-6880

Bahamas

Bahamas Tourist Office
150 East 52nd Street
New York, NY 10022
(212) 758-2777

Barbados

Barbados Board of Tourism
800 Second Avenue
New York, NY 10017
(800) 221-9831
(212) 986-6516

Belgium

Belgian Tourist Office
745 Fifth Avenue
New York, NY 10151
(212) 758-8130

Belize

Belize Tourist Board
415 Seventh Avenue, 18th Floor
New York, NY 10001
(800) 624-0686
(212) 268-8798

Bermuda

Bermuda Department of
Tourism
310 Madison Avenue
New York, NY 10017
(800) 223-6106

Bonaire

Bonaire Government
Tourist Office
201 1/2 East 29th Street
New York, NY 10016
(800) 826-6247

Tourism Corporation Bonaire
2039 Ninth Avenue
Ronkonkoma, NY 11779
(212) 779-0242

Brazil

Brazilian Consulate General
8484 Wilshire Boulevard,
Suites 730/711
Los Angeles, CA 90211
(213) 651-2664

British Virgin Islands

British Virgin Islands
Tourist Board
370 Lexington Avenue
New York, NY 10017
(212) 696-0400

Caribbean Region

Caribbean Tourism
Organization
20 East 46th Street, 4th Floor
New York, NY 10017
(212) 682-0435

TRAVEL INFORMATION FOR BIRDERS

Cayman Islands

Cayman Islands
Department of Tourism
420 Lexington Avenue,
Suite 2733
New York, NY 10170
(212) 682-5582

Chile

Chilean National
Tourist Board
510 West Sixth Street
Los Angeles, CA 90014
(213) 627-4293

China

China National Tourist Office
60 East 42nd Street, Suite 3126
New York, NY 10165
(212) 867-0271

Colombia

Colombia Government
Tourist Office
140 East 57th Street
New York, NY 10022
(212) 688-0151

Costa Rica

Costa Rican Tourist Board
3540 Wilshire Boulevard
Los Angeles, CA 90010
(213) 382-8080

Curaçao

Curaçao Tourist Board
400 Madison Avenue
New York, NY 10017
(212) 751-8266

Cyprus

Cyprus Consulate General
13 East 40th Street
New York, NY 10016
(212) 686-6016

Czechoslovakia

Czechoslovakia Cedok
10 East 40th Street, Suite 1902
New York, NY 10016
(212) 689-9720

Denmark

Danish Tourist Board
655 Third Avenue
New York, NY 10017
(212) 949-2333

Egypt

Egyptian Tourist Authority
323 Geary Street
San Francisco, CA 94102
(415) 781-7676

France

French Government
Tourist Office
610 Fifth Avenue
New York, NY 10020
(212) 757-1125

Germany

German National
Tourist Office
122 East 42nd Street
Chanin Building, 52nd Floor
New York, NY 10168-0072
(212) 661-7200

Travel and Tourism Offices

Great Britain

British Tourist Authority
40 West 57th Street
New York, NY 10019
(212) 581-4700

Greece

Greek National Tourist Office
645 Fifth Avenue
New York, NY 10022
(212) 421-5777

Grenada

Grenada Tourist Board
820 Second Avenue, Suite 9D
New York, NY 10017
(212) 687-9554

Honduras

Honduran Tourism Bureau
1138 Fremont Avenue
Los Angeles, CA 91030
(213) 682-3377

Hong Kong

Hong Kong Tourist Association
590 Fifth Avenue
New York, NY 10036-4706
(212) 869-5008

Iceland

Iceland Tourist Board
655 Third Avenue
New York, NY 10017
(212) 949-2333

India

Government of India
Tourist Office
3550 Wilshire Boulevard, Suite 204
Los Angeles, CA 90010-2485
(213) 380-8855

Government of India
Tourist Office
60 Bloor Street West, #1003
Toronto, Ontario M4W 3B8
(416) 962-3787

Indonesia

Indonesian Tourist Office
3457 Wilshire Boulevard
Los Angeles, CA 90010
(213) 387-2078
(213) 387-8309

Ireland

Irish Tourist Board
757 Third Avenue
New York, NY 10017
(212) 418-0800

Israel

Israeli Government
Tourist Office
350 Fifth Avenue
New York, NY 10018
(212) 560-0621

Italy

Italian Tourist Office
630 Fifth Avenue
New York, NY 10111
(212) 245-4961

Jamaica

Jamaica Tourist Board
801 Second Avenue, 20th Floor
New York, NY 10017
(212) 688-7650

Japan

Japan National Tourist Office
360 Post Street, Suite 401
San Francisco, CA 94108
(415) 989-7140

Japan National Tourist Office
2121 San Jacinto Street,
Suite 980
Dallas, TX 75201
(214) 754-1820

Kenya

Kenyan Tourist Office
424 Madison Avenue
New York, NY 10017
(212) 486-1300

Luxembourg

Luxembourg National
Tourist Office
801 Second Avenue
New York, NY 10017
(212) 370-9850

Macau

Macau Tourist Information
Bureau
Box 1860
3133 Lake Hollywood Drive
Los Angeles, CA 90078
(213) 851-3402
(800) 331-7150

Malaysia

Malaysian Tourist Centre
818 West Seventh Street
Los Angeles, CA 90017
(213) 689-9702

Mexico

Mexican Tourist Office
10100 Santa Monica
Boulevard, Suite 224
Los Angeles, CA 90067
(310) 203-8191

Monaco

Monaco Government
Tourist Office
845 Third Avenue
New York, NY 10022
(212) 759-5227

Morocco

Morocco National
Tourist Office
20 East 46th Street
New York, NY 10017
(212) 557-2520

The Netherlands

Netherlands Board of Tourism
225 North Michigan Avenue,
Suite 326
Chicago, IL 60601
(312) 819-0300

New Zealand

New Zealand Tourist Board
501 Santa Monica Boulevard
Los Angeles, CA 90055
(310) 477-8241

Travel and Tourism Offices

Norway

Norwegian Tourist Board
655 Third Avenue
New York, NY 10017
(212) 949-2333

Philippines

Philippine Consulate General
3660 Wilshire Boulevard,
Suite 216
Los Angeles, CA 90010
(213) 487-4525

Romania

Romanian National
Tourist Office
573 Third Avenue
New York, NY 10016
(212) 697-6971

Russia

Intourist Travel
Information Office
630 Fifth Avenue, Suite 868
New York, NY 10111
(212) 757-3884

Saba/St. Eustatius

Saba/St. Eustatius Tourism
Medhurst & Associates
271 Main Street
Northport, NY 11768
(516) 261-7474

Senegal

Senegal Tourist Office
888 Seventh Avenue, 27th Floor
New York, NY 10106
(212) 757-7115
(800) HI-DAKAR

South Africa

South African Tourism Board
747 Third Avenue
New York, NY 10017
(212) 838-8841
(800) 822-5368

Scandinavia

Scandinavia Tourist Boards
655 Third Avenue
New York, NY 10017
(212) 949-2333

Scotland

Scottish Tourist Board
551 Fifth Avenue
New York, NY 10176
(212) 986-2266

Spain

Spanish National
Tourist Office
665 Fifth Avenue
New York, NY 10022
(212) 759-8822

Spanish National
Tourist Office
102 Bloor Street West,
Suite 1850
Toronto, Ontario
(416) 961-3131

Sri Lanka

Sri Lankan Tourist Board
2148 Wyoming Avenue
Washington, DC 20008
(202) 483-4025

TRAVEL INFORMATION FOR BIRDERS

St. Kitts and Nevis

St. Kitts and Nevis Tourism
414 East 75th Street
New York, NY 10021
(212) 535-1234

St. Lucia

St. Lucia Tourist Board
820 Second Avenue, Suite 900E
New York, NY 10017
(212) 867-2950

St. Vincent and The Grenadines

St. Vincent and The
Grenadines Tourism
801 Second Avenue
New York, NY 10017
(212) 687-4981

Sweden

Swedish Tourist Board
655 Third Avenue
New York, NY 10017-5617
(212) 949-2333

Switzerland

Swiss National Tourist Office
608 Fifth Avenue
New York, NY 10020
(212) 757-5944

Taiwan

Taiwan Visitors Association
One World Trade Center
New York, NY 10048
(212) 466-0691

Thailand

Thailand Tourism Authority
3440 Wilshire Boulevard,
Suite 1100
Los Angeles, CA 90010
(213) 382-2353

Trinidad & Tobago

Trinidad & Tobago Tourism
Development Authority
25 West 43rd Street, Suite 1508
New York, NY 10036
(212) 719-0540
(800) 232-0082

Tunisia

Embassy of Tunisia
1515 Massachusetts
Avenue, NW
Washington, DC 20005
(202) 862-1850

Turks and Caicos

Turks and Caicos Tourism
The Keating Group
331 Madison Avenue
New York, NY 10017
(212) 888-4110

Turkey

Turkish Tourism Office
821 United National Plaza
New York, NY 10017
(212) 687-2194

Venezuela

Venezuelan Tourist Bureau
Seven East 51st Street
New York, NY 10022
(212) 826-1660

Chapter 3

BIRDERS ON TOUR

Birders in search of life birds—indeed, in search of any birds—love to travel. The skills of a birding tour operator can help make that travel simpler, more comfortable, and more successful in terms of birds seen. Tour operators generally have a lot of experience in their areas and have the valuable local knowledge of hotspots that can make the difference between a good and a great trip. Going with a tour operator need not be more expensive than going on your own. In fact, because of group rates, it may be cheaper.

Be selective when choosing a tour operator. Trips that meet every budget are available, but more expensive may not necessarily be better for you. Contact the tour operators for information and read it carefully. Ask about the size of the group and the level of birding expertise the other members are likely to have. Some birding tours are more oriented toward photographers than other—be sure to ask. And *always* ask for references to satisfied customers.

The list below is arranged by area of coverage—that is, by the region(s) in which the tour operator specializes. This can be slightly confusing, because a tour operator may specialize in trips to Alaska, for example, yet be based in some other state. Lodgings that cater to birders are also included, since these places can often arrange field trips and private guides for guests. Boat trips are in a separate section later in this chapter. *Winging It*, the monthly newsletter of the American Birding Association, publishes detailed annual listings of tour operators and pelagic trips. In addition, each issue of *Winging It* contains a listing of birding tours sponsored by the ABA because of their unique destinations and excellent leaders.

Additional information about guided bird tours worldwide is available as moderately priced publications from:

>Guides to Guided Bird Tours
>27 Tall Cedar Court
>RR 6
>Belle Mead, NJ 08502
>(908) 359-2097

Traveling birders can find information about reasonably priced overnight accommodations with local birdwatchers from:

>Birds of a Feather
>Tony and Sherry Zoars
>Box 915
>La Grange, IL 60525
>(708) 354-2559

In the unlikely event that you have a serious complaint or problem with a birding tour operator, contact your state's consumer protection agency and:

American Society of Travel Agents
1101 King Street, Suite 200
Alexandria, VA 22314
(703) 739-2782

or

United States Tour Operators Association
211 East 51st Street, Suite 12B
New York, NY 10022
(212) 944-5727

TOUR OPERATORS—UNITED STATES

Alaska

Afognak Wilderness Lodge
Box 1
Seal Bay, AK 99697
(907) 486-6442

Alaska Maritime Tours
Box 3098
Homer, AK 99603
(907) 235-2490

Alaska Up Close
Box 32666
Juneau, AK 99803
(907) 789-9544

Alaska Wild Wings
Goose Cove Lodge
Box 325
Cordova, AK 99574
(907) 424-5111

Attour, Inc.
2027 Partridge Lane
Highland Park, IL 60035
(708) 831-0207

Borderland Tours
2550 West Calle Padilla
Tucson, AZ 85745
(800) 525-7753
(602) 882-7650

Canadian River Expeditions
424-845 Chilco Street
Vancouver, British Columbia
V6G 2R2
(604) 738-4449

Center for Alaska
Coastal Studies
Box 2225
Homer, AK 99603
(907) 235-6667

Cheeseman's Ecology Safaris
20800 Kittredge Road
Saratoga, CA 95070
(800) 527-5330

Harper and Hallet
697 Darlington Road NE
Atlanta, GA 30305
(404) 233-3974

International Expeditions, Inc.
One Environs Park
Helena, AL 35080
(800) 633-4734
(205) 428-1700

Kachemak Bay
Wilderness Lodge
China Poot Bay
Box 956
Homer, AK 99003
(907) 235-8910

Joseph Van Os Photo Safaris
and Nature Tours
Box 655
Vashon Island, WA 98070
(800) 368-0077

Wilderness Birding Adventures
Bob Dittrick
Box 10-3747
Anchorage, AK 99510
(907) 694-7442

BIRDERS ON TOUR

Arizona

AdvenTours
Box 1441
Longmont, CO 80502
(303) 440-8882

Best of Nature
David A. Jasper
Box 430
Portal, AZ 85632
(602) 558-2307

Borderland Tours
2550 West Calle Padilla
Tucson, AZ 85745
(800) 525-7753
(602) 882-7650

Cal Nature Tours
SVL 7310
Victorville, CA 92392
(619) 241-2322

Casa Alegre Bed &
Breakfast Inn
316 East Speedway
Tucson, AZ 85705
(602) 628-1800

Circle Z Guest Ranch
Box 194W
Patagonia, AZ 85624
(602) 287-2091

Penfeathers Tours
Box 38157
Houston, TX 77238
(713) 445-1187

Portal Bed and Breakfast
Box 364
Portal, AZ 85632
(602) 558-2223

Quality Inn Green Valley
111 South La Cañada
Green Valley, AZ 85614
(800) 344-1441
(602) 625-2250

Ramsey Canyon Inn
Bed and Breakfast
31 Ramsey Canyon Road
Hereford, AZ 85615
(602) 378-3010

Santa Rita Lodge and
Nature Resort
HC 70, Box 5444
Sahuarita, AZ 85629
(602) 625-8746

Sarett Nature Center
2300 Benton Center Road
Benton Harbor, MI 49022
(616) 927-4832

The Nature Conservancy,
Aravaipa Canyon Preserve
Klondyke Station
Willcox, AZ 85643
(602) 828-3443

The Nature Conservancy,
Mile-Hi and Ramsey Canyon
Preserve
RR 1, Box 84
Hereford, AZ 85615
(602) 378-2785

Tour Operators—United States

The Nature Conservancy,
Muleshoe Ranch
RR 1, Box 1542
Willcox, AZ 85643
(602) 749 5555

Arkansas

Lake Chicot State Park
Route 1, Box 1555
Lake Village, AR 71653
(800) 264-2430
(501) 265-5480

California

Borderland Productions
2550 West Calle Padilla
Tucson, AZ 85745
(800) 525-7753
(602) 882-7650

Martin Byhower
1628 Armour Lane
Redondo Beach, CA 90278
(310) 374-7473

Cal Nature Tours
SVL 7310
Victorville, CA 92392
(619) 241-2322

Redwood Coast Birdwatching
Box 670
Crescent City, CA 95531

The Nature Conservancy
3152 Paradise Drive
Tiburon, CA 94920
(415) 281-0423

Colorado

Penfeathers Tours
Box 38157
Houston, TX 77238
(713) 445-1187

Sage Grouse Ranch
Box 313
Gunnison, CO 81230
(303) 641-0793

Joseph Van Os Photo Safaris
and Nature Tours
Box 655
Vashon Island, WA 98070
(800) 368-0077

Florida

Mort Cooper
7625 SE 97th Court
Miami, FL 33173
(305) 271-2413

Fred Dayhoff,
Wilderness Guide
SR Box 68
Ochopee, FL 33943
(813) 695-4360

Florida Nature Tours
Box 5643
Winter Park, FL 32793
(407) 273-4400

Jeff Goodwin Nature Tours
6745 SW 132nd Avenue,
Suite 211
Miami, FL 33183
(305) 388-7430

Overton Ganong Southeast
Nature Tours
Box 320291
Tampa, FL 33679
(813) 289-0891

Joseph Van Os Photo Safaris
and Nature Tours
Box 655
Vashon Island, WA 98070
(800) 368-0077

Noel Wamer
502 East Georgia Street
Tallahassee, FL 32303
(904) 222-2084

Georgia

Environmental Concern Inc.
Box P
St. Michaels, MD 21663
(301) 745-9620

Little St. Simons Island
Box 1078
St. Simons Island, GA 31522
(912) 638-7472

Hawaii

David Kuhn
Box 1018
Waimea, HI 96796
(808) 335-3313

Idaho

Raptours
Box 1191
Annandale, VA 22003
(703) 642-2386

Illinois

Wilderness Institute
Box 547
Lisle, IL 60532
(800) 842-0207

Louisiana

Honey Island Swamp Tours
106 Holly Ridge Drive
Slidell, LA 70461
(504) 641-1769

Massachusetts

Berkshire Hiking Holidays
Box 2232
Lenox, MA 02140
(413) 499-9648

Bill Drummond
24 Elm Street
Haverhill, MA 01830
(508) 373-4292

Michigan

Budget Birding
1731 Hatcher Crescent
Ann Arbor, MI 48103
(313) 995-4357

Great Lakes Birding
303 Kenwood
Roscommon, MI 48653
(517) 257-2161

NettieBay Lodge
8811 West 638 Highway
Hawks, MI 49743
(517) 734-4688

Tour Operators—United States

Quinss Inn
3751 North Bell
Chicago, IL 60618
(312) 248-2817

Raptours
Box 1191
Annandale, VA 22003
(703) 642-2386

Sarett Nature Center
2300 Benton Center Road
Benton Harbor, MI 49022
(616) 927-4832

Wilderness Institute
Box 547
Lisle, IL 60532
(800) 842-0207

Minnesota

Kim Eckert
8255 Congdon Boulevard
Duluth, MN 55804
(218) 525-6930

National Forest Lodge
3226 Highway One
Isabella, MN 55607
(218) 323-7676

Wild Trails
Box 34
Spring Lake, MN 56680
(218) 789-2639

Mississippi

Mississippi Coast Birding
Four Hartford Place
Gulfport, MS 39507
(601) 886-3153

Tara Wildlife
6791 Eagle Lake Shores Road
Vicksburg, MS 39180
(601) 279-4261

Montana

Bitterroot River Lodge
Box 1248
Hamilton, MT 59840
(406) 363-5191

Centennial Birding Tours
Box 741
Dillon, MT 59725

The Nature Conservancy
Montana Field Office
Box 258
Helena, MT 59624
(406) 443-0303

New Jersey

Blue Amber Motel
605 Madison Avenue
Cape May, NJ 08204
(609) 884-8266

Cape May Bird Observatory
Box 3
Cape May Point, NJ 08212
(609) 884-2736

New Mexico

Bear Mountain Guest Ranch
Box 1163
Silver City, NM 88062
(505) 538-2538

Eaton House Bed and Breakfast
Box 536
Socorro, NM 87801
(505) 835-1067

Star Hill Inn
Sapello, NM 87745
(505) 425-5605

New York

Adirondack Bird Seminar
RD 1, Box 85
Ghent, NY 12075
(508) 828-7007

Waterman Conservation
Education Center
403 Hilton Road
Apalachin, NY 13732
(607) 625-2221

North Carolina

FENCE (Foothills Equestrian
Nature Center)
500 Hunting Country Road
Tryon, NC 28782

Brian Patteson
Box 125
Amherst, VA 24521
(804) 933-4456

North Dakota

Dakota Bird Tours
Box 605
Clear Lake, SD 57226
(605) 874-2223

Oregon

Malheur Field Station
HC 72, Box 260
Princeton, OR 97721
(503) 493-2629

SAS Tours
Seattle Audubon Society
8028 35th Avenue NE
Seattle, WA 98115

The Nature Conservancy
Oregon Field Office
1205 NW 25th Avenue
Portland, OR 97210
(503) 228-9561

Rhode Island

Block Island Birding Weekend
Audubon Society of
Rhode Island
12 Sanderson Road
Smithfield, RI 02917
(401) 231-6444

South Carolina

FENCE (Foothills Equestrian
Nature Center)
500 Hunting Country Road
Tryon, NC 28782

South Dakota

Dakota Bird Tours
Box 605
Clear Lake, SD 57226
(605) 874-2223

Tour Operators—United States

Tennessee

Appalachian Birding Weekends
114 Malone Hollow Road
Jonesborough, TN 37659
(615) 753-7831

Texas

Big Bend Birding Expeditions
Box 507
Terlingua, TX 79852
(915) 371-2356

Eagle Nest Ranch
HCR 1, Box 44
Barksdale, TX 78828
(512) 483-5425

Matagorda Island Birding
Diane Lillard
Matagorda Island, TX 77457
(512) 877-7024

Penfeathers Tours
Box 38157
Houston, TX 77238
(713) 445-1187

Raptours
Box 1191
Annandale, VA 22003
(703) 642-2386

Sea-Gun Resort
Highway 35 North and Park Road 13
Fulton, TX 78358
(512) 729-3292

Wilderness Institute
Box 547
Lisle, IL 60532
(800) 842-0207

Utah

Colorado River and Trail Expeditions, Inc.
Box 57575
Salt Lake City, UT 84157
(800) 253-7328

Learning Adventures
1556B Georgia
Boulder City, NV 89005
(702) 293-5758

Virginia

Burton House Bed & Breakfast
Box 182
Wachapreague, VA 23480
(804) 787-4560

Wachapreague Motel and Restaurant
Box 360
Wachapreague, VA 23480
(804) 787-2105

Washington

Robert Ashbaugh
Box 84205
Seattle, WA 98124
(206) 823-2219

Raven Kayak Experiences
Box 5303
Bellingham, WA 98227
(206) 671-4528

BIRDERS ON TOUR

SAS Tours
Seattle Audubon Society
8028 35th Avenue NE
Seattle, WA 98115

West Virginia

Oglebay Institute
Brooks Nature Center
Wheeling, WV 26003
(304) 242-6855

Wyoming

Yellowstone Wildlife Tours
Box 546
Yellowstone National Park,
WY 82190
(406) 848-7942

TOUR OPERATORS—CANADA

All Canada

Canadian Nature Tours
355 Lesmill Road
Don Mills, Ontario M3B 2W8
(416) 444-8419

Alberta

Birdwood Bed & Breakfast
RR 2
Fort Saskatchewan, Alberta
T8L 2NB
(403) 998-0082

British Columbia

Canadian River Expeditions
424-845 Chilco Street
Vancouver, British Columbia
V6G 2R2
(604) 738-4449

Merlin House Bed and
Breakfast
3762 West First Avenue
Vancouver, British Columbia
V6R 1H2
(604) 736-9471

Manitoba

Churchill Wilderness
Encounter
Box 9
Churchill, Manitoba R0B 0E0
(204) 675-2248

Mort Cooper
7625 SE 97th Court
Miami, FL 33173
(305) 271-2413

Joseph Van Os Photo Safaris
and Nature Tours
Box 655
Vashon Island, WA 98070
(800) 368-0077

New Brunswick

Shorecrest
North Head
New Brunswick E0G 2M0
(506) 662-3216

Wonder Bird Tours, Inc.
Box 2015
New York, NY 10159
(800) BIRDTUR
(212) 736-BIRD

Tour Operators—Canada

Northwest Territories

Arctic Waterways, Inc.
RR 2
Stevensville, Ontario L0S 1S0
(416) 382-3882

Bathhurst Inlet Lodge
Box 820
Yellowknife, Northwest
Territories X1A 2N6
(403) 873-2595

Oldsquaw Lodge
Bag Service 2711
Whitehorse, Yukon Y1A 3V5
(403) 668-6732

Nova Scotia

Wonderbird Tours, Inc.
Box 2015
New York, NY 10159
(800) 247-3887
(212) 727-0780

Ontario

Budget Birding
1731 Hatcher Crescent
Ann Arbor, MI 48103
(313) 995-4357

Creeway Wilderness
Experiences
Box 347G
Moose Factory, Ontario P0L
1W0
(705) 658-4390

Federation of Ontario
Naturalists
428 Falconer Street
Port Elgin, Ontario N0H 2C2

Flora and Fauna Fieldtours
232 Bolton Drive
Bolton, Ontario L7E 1Z7
(416) 857-2235

Point Pelee Birding
Alan Wormington
RR 1
Leamington, Ontario N8H 3V4
(519) 326-0687

Polar Bear Lodge
Moosonee, Ontario P0L 1Y0
(705) 336-2345
(416) 244-1495 (Toronto)

Quebec

Wonderbird Tours, Inc.
Box 2015
New York, NY 10159
(800) 247-3887
(212) 727-0780

Yukon Territory

Canadian River Expeditions
424-845 Chilco Street
Vancouver, British Columbia
V6G 2R2
(604) 738-4449

Oldsquaw Lodge
Bag Service 2711
Whitehorse, Yukon Y1A 3V5
(403) 668-6732

North America

Arete Tours, Inc.
Box 620362
Middleton, WI 53562
(800) 236-BIRD

Bird Bonanzas
Box 611563
North Miami, FL 33161
(305) 895-0607

Bill Drummond Birding Tours
24 Elm Street
Haverhill, MA 01830
(508) 373-4292

Ben Feltner's Peregrine Tours
Box 4251
Seattle, WA 98106
(206) 767-9937

FOCUS Nature Tours
Box 21230
St. Petersburg, FL 33742
(813) 522-3338

Field Guides Incorporated
Box 160723
Austin, TX 78716
(512) 327-4953

Flora and Fauna Field Tours
232 Bellair Drive
Bolton, Ontario L7E 1Z7
(416) 857-2235

Focus on Nature Tours
(FONT)
Box 9021
Wilmington, DE 19809
(800) 547-1070
(302) 529-1876

Goldeneye Nature Tours
Box 30416
Flagstaff, AZ 86003
(800) 624-6606

Clive and Joy Goodwin
Enterprises, Ltd.
One Queen Street, Suite 401
Cobourg, Ontario K9A 1M6
(416) 372-1065

Dan Guravich's Select Tours
Photo Safaris
Box 891
Greeneville, MS 38701
(601) 335-2414

Harper and Hallet
697 Darlington Road NE
Atlanta, GA 30305
(404) 233-3974

Massachusetts Audubon
Society
Natural History Travel
South Great Road
Lincoln, MA 01173
(800) 289-9504
(617) 259-9500

Miller's Nature Tours
RR 1
Annan, Ontario N0H 1B0
(519) 376-6366

MotMot Nature Tours
1018 West Upland Road
Ithaca, NY 14850
(607) 257-6686

Tour Operators—International

National Audubon Society
700 Broadway
New York, NY 10003
(212) 979-3000

Nature Travel Service
127B Princess
Kingston, Ontario K7L 1A8
(613) 546-3065

OBServ Tours
3901 Trimble Road
Nashville, TN 37215
(615) 292-2739

Overton Ganong-Southeast
Nature Tours
Box 320291
Tampa, FL 33679
(813) 289-0891

Parula Tours
1711 West Oglethorpe Avenue
Albany, GA 31707

Questers
257 Park Avenue South
New York, NY 10010
(800) 468-8668
(212) 673-3120

Dennis Smeltzer
826 Sewickly Street
Greensburg, PA 15601

Victor Emanuel Nature Tours
(VENT)
Box 33008
Austin, TX 78764
(800) 328-VENT
(512) 328-5221

WINGS
Box 31930
Tucson, AZ 85751
(602) 749-1967

TOUR OPERATORS—INTERNATIONAL

Africa

Cheeseman's Ecology Safaris
20800 Kittredge Road
Saratoga, CA 95070
(800) 527-5330

Asia

KingBird Tours
Box 196
Planetarium Station
New York, NY 10024
(212) 866-7923

Australia

Cassowary House
Kuranda, Queensland 4872
Australia

Diomedea Pty. Ltd.
Alan McBride
Box 190
Cremorne Junction,
New South Wales 2090
Australia
61-2-953-2546

Emu Tours
Box 4
Jameroo, New South Wales
2533
Australia

Falcon Tours
One Simons Drive
Roleystone, Western Australia
6111
Australia

Gipsy Point Lodge
Alan Robertson
Gipsy Point, Victoria 3891
Australia
051-58-8205

Inland Bird Tours
94 Hunter Street
Deniliquin, New South Wales
2710
Australia
058-815278

Kirrama Wildlife Tours
Box 133
Silkwood, North Queensland
4856
Australia
61-07-655197 (fax)

Kingfisher Park
Box 3
Julatten, Queensland 4871
Australia

Philip Maher
94 Hunter Street
Deniliquin, New South Wales
2710
Australia
058-813378

Graham Pizzey
Victoria Valley Road
Dunkeld, Victoria 3191
Australia

Belize

Chan Chich Lodge
Box 37
Belize City, Belize
(800) 343-8009
(501) 2-75634

Central and South America

Borderland Tours
2550 West Calle Padilla
Tucson, AZ 85745
(800) 525-7753

Condor Pacific Eco Tours of
the Americas
1730 Omie Way
Lawrenceville, GA 30243
(404) 995-7537

Flora and Fauna Fieldtours
232 Bellair Drive
Bolton, Ontario L7E 1Z7
(416) 857-2235

Focus on Nature Tours
(FONT)
Box 9021
Wilmington, DE 19809
(800) 547-1070
(302) 529-1876

Neotropic Bird Tours
38 Brookside Avenue
Livingston, NJ 07039
(800) 662-4852

Osprey Tours
Box 832
West Tisbury, MA 02575
(508) 645-9049

Tour Operators—International

Costa Rica

Birdwatch Costa Rica
Box 025216
Miami, FL 33102

Borderland Tours
2550 West Calle Padilla
Tucson, AZ 85745
(800) 525-7753

Olga L. Clarke
2027 El Arbolita Drive
Glendale, CA 91208
(818) 249-5537

Costa Rica Resources Co.
10031 Fourth Avenue, Suite 3G
Bay Ridge, NY 11209
(718) 748-2158

Miller Nature Tours
RD 1, Box 1152
Maryland, NY 12116
(607) 432-5767

Organization for
Tropical Studies
Box 90630
Durham, NC 27708
(919) 684-5774

La Posada de la Montaña
Costa Rica
Box 308
Greenfield, MO 65661
(800) 632-3892

Rancho Naturalista
Apartado 364-1002
San José, Costa Rica
(506) 39-7138

Selva Verde Lodge
c/o Costa Rica Experts
3540 NW 13th Street
Gainesville, FL 32609
(800) 858-0999

Michael Snow
Apartado 73
7200 Siquirres
Costa Rica

Las Ventana de Osa
Wildlife Refuge
Box 1089
Lake Helen, FL 32744
(904) 228-3356

Ecuador

La Selva Jungle Lodge
6 de Diciembre 2816
Quito, Ecuador
550-995
554-686

Galápagos

Galápagos Travel
Box 1220
San Juan Batista, CA 95045
(800) 969-9014

Hong Kong

Kingfisher Tours
Two Villa Paloma
Shuen Wan, Tai Po
Hong Kong
(852) 665-8190

Ireland

Irish Ecology Tours
Richard Mine Road, Suite E8
Wharton, NJ 07885
(800) 242-7020

Jamaica

Arete Tours
Box 362
Middleton, WI 53563
(608) 831-8235

The Touring Society
of Jamaica
Box 13
Duncans, Jamaica
(800) 624-4935
(809) 925-2253

Russia

K & I Excursions
18201 Evergreen
Villa Park, CA 92667
(800) 285-9845

Scandinavia

Destinations
114 Malone Hollow Road
Jonesborough, TN 37659
(615) 753-7831

South America

Condor Pacific Tours
1730 Omie Way
Lawrenceville, GA 30243
(800) 783-8847
(404) 995-7537

Neotropic Bird Tours
38 Brookside Avenue
Livingston, NJ 07039
(800) 662-4852

Trinidad & Tobago

Peregrine Enterprises
Bill Murphy
101 Rathbone Terrace
Marietta, OH 45750
(614) 373-3966

Asa Wright Nature
Centre and Lodge
c/o Caligo Ventures, Inc.
156 Bedford Road
Armonk, NY 10504
(800) 426-7781

United Kingdom

Avian Adventures
Three Woodhaven
Wedges Mills, Cannock
WS11 1RE
England
922-417102

Borrobol Birding (Scotland)
c/o Josephine Barr
(800) 323-5463
(708) 251-4110

Bob Buckler
Three Kings Road
Sherborne, Dorset DT9 4HU
England
01144-935-816176

Lapwing Tours
2836 Patilla Avenue
Vero Beach, FL 32960
(407) 562-5247

Tour Operators—International

Nick Pope
2/38 Carshalton Grove
Sutton, Surrey SM1 4LZ
England
4481-661-0401

Wessex Bird Tours
12 Redland Court Road
Bristol BS6 7EQ
Avon, England
01144-272-246255

Venezuela

Cheeseman's Ecology Safaris
20800 Kittredge Road
Saratoga, CA 95070
(800) 527-5330

International

Abercrombie & Kent
1529 Kensington Road
Oak Brook, IL 60521
(800) 323-7308

Birdquest
Two Jays, Kemple End
Birdy Brow, Stonyhurst
Lancashire BB6 9QY
England
44-254-826317

Birds and Birders
Box 737
9700 AS Groningen
Holland

Borderland Tours
2550 West Calle Padilla
Tucson, AZ 85745
(800) 525-7753

Cal Nature Tours
7310 SVL Box
Victorville, CA 92392
(619) 241-2322

Cheeseman's Ecology Safaris
20800 Kittredge Road
Saratoga, CA 95070
(800) 527-5330

Clipper Adventure Cruises
7711 Bonhomme Avenue
St. Louis, MO 63015
(800) 325-0010
(314) 727-6576

Destinations
114 Malone Hollow Road
Jonesborough, TN 37659
(615) 753-7831

Eco-Expeditions
1414 Dexter Avenue North,
Suite 327
Seattle, WA 98109
(800) 628-8747

Explorer Shipping Corporation
1520 Kensington Road,
Suite 201
Oak Brook, IL 60521
(800) 323-7308

Focus on Nature Tours
(FONT)
Box 9021
Wilmington, DE 19809
(800) 547-1070
(302) 529-1876

Holbrook Travel, Inc.
3540 NW 13th Street
Gainesville, FL 32609
(800) 451-7111
(904) 377-7111

International Expeditions Inc.
One Environs Park
Helena, AL 35080
(800) 633-4734
(205) 429-1700

Martin Travel
5216 Pershing Avenue
Fort Worth, TX 76107
(817) 377-BIRD

National Audubon Society
700 Broadway
New York, NY 10003
(212) 979-3000

Peregrine Enterprises Inc.
101 Rathbone Terrace
Marietta, OH 45750
(614) 373-3966

Princeton Nature Tours, Inc.
282 Western Way
Princeton, NJ 08540
(609) 683-1111

Questers Worldwide
Nature Tours
257 Park Avenue South
New York, NY 10010
(800) 468-8668

Voyagers International
Dave Blanton
Box 915
Ithaca, NY 14851
(800) 633-0299

Wild Goose Travel
Box 706
Pacifica, CA 94044
(800) 432-2391

Woodstar Tours, Inc.
908 South Massachusetts
Avenue
De Land, FL 32724
(904) 736-0327

Zegrahm Expeditions
1414 Dexter Avenue North,
Suite 327
Seattle, WA 98109
(800) 628-8747

BIRDING BY BOAT

Going on a boat trip is one of the most enjoyable birding adventures—and a great way to fill in all those blanks at the start of your life list. Most but not all of the trips listed below are pelagic (oceangoing) voyages designed to find shearwaters, albatrosses, and other offshore birds. A pelagic trip is generally an all-day affair, since it often involves sailing well out into the ocean to find the birds. Shorter boat trips to see puffins, whooping cranes, and the like are also listed below. Schedules vary depending on the season and weather. Always write or call well in advance.

Alaska

Alaska Marine
Highway System
Box 25535
Juneau, AK 99802
(800) 642-0066
(907) 465-3946

Alaska Maritime Tours, Inc.
Box 3098
Homer, AK 99603
(907) 235-2490

Kenai Coastal Tours, Inc.
524 West Fourth Avenue
Anchorage, AK 99501
(800) 770-9119
(907) 277-2131

Kenai Fjords Tours, Inc.
Box 1889
Seward, AK 99664
(907) 224-8068

Mariah Charters
3812 Katmai Circle
Anchorage, AK 99517
(907) 224-8623 (summer)
(907) 243-1238 (winter)

Snow Bird Charters
Box 1066
Seward, AK 99664
(907) 224-5582

California

Gail and Doug Cheeseman
20800 Kittredge Road
Saratoga, CA 95070
(408) 867-1371

Clipper Cruise Line
Windsor Building
7711 Bonhomme Avenue
St. Louis, MO 63105
(800) 635-5062
(314) 727-2929

Goldeneye Nature Tours
John Shipley, Jr.
Box 20755
Mesa, AZ 85277
(602) 962-4303

Los Angeles Audubon Society
Phil Sayre
7377 Santa Monica Boulevard
Los Angeles, CA 90046
(213) 876-0202

Oceanic Society Expeditions
Fort Mason Center, Building E
San Francisco, CA 94129
(415) 474-3385

Shearwater Journeys
Debra Love Shearwater
Box 1445
Soquel, CA 95073
(408) 688-1990

Western Field Ornithologists
Stan Walens
4411 Huggins Street
San Diego, CA 92122
(619) 450-0258

WINGS, Inc.
Box 31930
Tucson, AZ 85751
(602) 749-1967

Delaware

Focus on Nature Tours
Box 9021
Wilmington, DE 19809
(302) 529-1876

Florida

Field Guides, Inc.
Box 160723
Austin, TX 78716
(512) 327-4953

Florida Nature Tours
Wes Biggs
Box 5643
Winter Park, FL 32793
(407) 273-4400

WINGS, Inc.
Box 31930
Tucson, AZ 85751
(602) 749-1967

Yankee Tortugas Safaris
Box 5903
Key West, FL 33040
(800) 634-0939

Georgia

Georgia Ornithological Society
c/o Bill Blakeslee
1722 Noble Drive NE
Atlanta, GA 30306
(404) 881-6570

Maine

Atlantic Expeditions
HCR 35, Box 290
St. George, ME 04857
(207) 372-8621

Bluenose Ferry
Marine Atlantic Co.
Box 250
North Sydney, Nova Scotia
B2A 3M3
(800) 341-7981
(902) 794-5700

Machias Seal Island Trips
Capt. John Norton
RR 1, Box 990
Jonesport, ME 04649
(207) 497-5933

Matinicus Rock Boat Trips
Matinicus, ME 04851
(207) 366-3700
(207) 366-3830

Captain Andrew Patterson
Box 364
Cutler, ME 04626
(207) 259-4484

WINGS, Inc.
Box 31930
Tucson, AZ 85751
(602) 749-1967

Maryland

Atlantic Seabirds
Gene Scarpulla
7906-B Knollwood Road
Towson, MD 21286
(410) 821-0575

Brian Patteson, Inc.
Box 1135
Amherst, VA 24521
(804) 933-8687

Virginia Pelagic Trips
Ken Bass
Box 572
Nokesville, VA 22123
(703) 594-2714

Massachusetts

Bruce Hallett
697 Darlington Road
Atlanta, GA 30305
(404) 261-4322

Massachusetts Audubon
Society
Wellfleet Bay Wildlife
Sanctuary
Box 236
Wellfleet, MA 02663
(508) 349-2615

New Jersey

Alan Brady
Box 103
Wycombe, PA 18980
(215) 598-7856

Jersey Cape Nature Excursions
Box 254
Cape May, NJ 08204
(609) 898-9631

New Hampshire

Isles of Shoals Steamship Co.
Box 311
315 Market Street
Portsmouth, NH 03802
(800) 441-4620
(603) 431-5500

North Carolina

Brian Patteson, Inc.
Box 1135
Amherst, VA 24521
(804) 933-8687

Focus on Nature Tours
Box 9021
Wilmington, DE 19809
(302) 529-1876

OBServ Tours
Bob Odear
3901 Trimble Road
Nashville, TN 37215
(615) 292-2739

Pterodroma Ptours
Michael Tove
303 Dunhagen Place
Cary, NC 27511
(919) 460-0338

Oregon

Portland Oregon Society
Audubon House
5151 NW Cornell Road
Portland, OR 97210
(503) 292-6855

Rhode Island

Charles Avenengo
Box 3724
Peace Dale, RI 02883
(800) 662-2824

South Carolina

Chesapeake Bay Nature
Cruises and Expeditions
Box 833
St. Michaels, MD 21663
(800) 344-3255

Texas

Captain Ted's Whooping
Crane Tours
HCO 1, Box 225J
Rockport, TX 78382
(800) 338-4551

Fisherman's Wharf
Box 387
Port Aransas, TX 78373
(512) 749-5760

Ray Little
Box 387
Port Aransas, TX 78373
(512) 749-5760

Whooping Crane Tours
Lucky Day Charters
1903 Glass Avenue
Rockport, TX 78382
(512) 729-GULL
(512) 782-BIRD

Virginia

Chesapeake Bay Nature
Cruises and Expeditions
Box 833
St. Michaels, MD 21663
(800) 344-3255

Brian Patteson, Inc.
Box 1135
Amherst, VA 24521
(804) 933-8687

Virginia Pelagic Trips
Ken Bass
Box 572
Nokesville, VA 22123
(703) 594-2714

Washington

Black Ball Transport, Inc.
10777 Main, Suite 106
Bellevue, WA 98004
(206) 622-2222

Portland Oregon Society
Audubon House
5151 NW Cornell Road
Portland, OR 97210
(503) 292-6855

Terry Wahl
3041 Eldridge
Bellingham, WA 98225
(206) 733-8255

Washington State Ferry System
801 Alaskan Way
Seattle, WA 98104
(800) 542-0810
(206) 464-4600

CANADA

Atlantic Maritime Provinces Ferries

Bluenose Ferry
Marine Atlantic Co.
Box 250

Birding by Boat

North Sydney, Nova Scotia
B2A 3M3
(800) 341-7981
(902) 794-5700

Goose Bay Ferry,
Labrador/Newfoundland
Marine Atlantic Co.
Box 250
North Sydney, Nova Scotia
B2A 3M3
(800) 341-7981
(902) 794-5700

Argentia Ferry,
Newfoundland/Nova Scotia
Marine Atlantic Co.
Box 250
North Sydney, Nova Scotia
B2A 3M3
(800) 341-7981
(902) 794-5700

Port-aux-Basques Ferry,
Newfoundland/Nova Scotia
Marine Atlantic Co.
Box 250
North Sydney, Nova Scotia
B2A 3M3
(800) 341-7981
(902) 794-5700

British Columbia

Black Ball Transport, Inc.
10777 Main, Suite 106
Bellevue, WA 98004
(206) 622-2222

British Columbia Ferry Corp.
1112 Fort Street
Victoria, British Columbia
V8V 4V2
(604) 386-3431

Kallahin Travel Services
Box 131D
Queen Charlotte City, British
Columbia V0T 1S0
(604) 559-8455

Seasmoke Sailing
Charters and Tours
Box 483
Alert Bay, British Columbia
V0N 1A0
(604) 974-5225

Washington State Ferry System
801 Alaskan Way
Seattle, WA 98104
(800) 542-0810
(206) 464-4600

New Brunswick

Goldeneye Nature Tours
Box 20755
Mesa, AZ 85277
(602) 962-4303

Island Coast Boat Tours
Castalia, Grand Manan, New
Brunswick E0G 1L0
(506) 662-8181

Nova Scotia

Brier Island Whale and Seabird
Cruises, Ltd.
Westport, Nova Scotia
B0V 1H0
(902) 839-2995

BIRDING EVENTS

Many birders plan their December around the venerable Christmas Bird Count. There are many other interesting and educational annual birding events. Dates vary somewhat from year to year; always check well in advance.

The Christmas Bird Count

The Christmas Bird Count began on Christmas Day in 1900, when Frank Chapman organized 25 groups in the Northeast as a protest against the traditional holiday slaughter in which teams competed to see who could shoot the most birds in one day. Today the National Audubon Society sponsors its annual Christmas Bird Count every year from December 17 to January 3. This event enlists the aid of over 41,000 bird watchers in every state, every province of Canada, and several countries in South America, Central America, and the Caribbean islands. Participants spend an entire day, regardless of weather, censusing the birds in their region. Every bird that can be identified is listed and later reported back to Audubon. Every year some 1,200 different species are identified, and well over 74 million individual birds are counted by over 1,500 groups.

The data provide critical information about the winter distribution of resident bird species and allow researchers to track changes in bird populations and ranges. The Christmas Bird Count is organized by *American Birds*, which devotes a book-sized issue every year to analyzing the results.

Any birder wishing to participate in a Christmas Bird Count should contact a local Audubon Society chapter or write to the Christmas Bird Count Editor at *American Birds*. Many counts are held on a Saturday or Sunday; a modest fee of a few dollars is collected to help defray expenses. Each count area is a unique circle 15 miles (24 km) in diameter, encompassing about 175 square miles (455 km). Groups try to cover as much territory within the circle as is possible within the 24 hours of a single calendar day. Not surprisingly, some groups devote weeks to planning the logistics in order to claim the longest list of birds. The record is 326, noted in the

Birding Events

Atlantic Canal Area of Panama in 1986. Some counters have been known to travel by dogsled to find their birds.

National

North American Migration Count
Jim Stasz
Box 71
North Beach, MD 20714

Project FeederWatch
Cornell Laboratory of Ornithology
159 Sapsucker Woods Road
Ithaca, NY 14850
(607) 254-BIRD

UNITED STATES

Alaska

Copper River Delta Shorebird Festival
Cordova Chamber of Commerce
Box 99
Cordova, AK 99574
(907) 424-7260
Date: early May

Kachemak Bay Shorebird Festival
Homer Chamber of Commerce
Box 541
Homer, AK 99603
(907) 235-7740
Date: early May

Arizona

Southwest Wings Birding Festival
Sierra Vista Chamber of Commerce
77 Calle Portal
Sierra Vista, AZ 85635
(800) 288-3861
(602) 458-6940
Date: late August

California

Annual Bald Eagle Conference
Klamath Basin National Wildlife Refuges
Route 1, Box 74
Tulelake, CA 96134
(916) 667-2231
Date: winter

Georgia

Weekend for Wildlife
Georgia Department of Natural Resources
205 Butler Street SW, Suite 1258
Atlanta, GA 30334
(404) 656-0772
Date: spring

Iowa

Bald Eagle Appreciation Days
Keokuk Area Convention and Visitors Bureau
401 Main Street
Keokuk, IA 52632
(319) 524-5055
Date: January

Michigan

District Ranger,
U.S. Forest Service
Huron National Forest
Mio, MI 48652
(517) 826-3252
Dates: late May, June

Midwest Birding Symposium
2300 Benton Center Road
Benton Harbor, MI 49022
(616) 927-4832
Date: September

Nebraska

Crane Watch
Keerney Visitors Bureau
Box 607
Keerney, NE 68948
(800) 227-8340
Date: February through April

Wings Over the Platte
Hall County Conservation
and Visitors Bureau
Box 1482
Grand Island, NE 68802
(800) 658-3178

New Jersey

Cape May Weekends
Cape May Bird Observatory
Box 3
Cape May Point, NJ 08212
(609) 884-2736
Dates: spring, autumn

Pine Barrens Weekend
Rancocas Nature Center
New Jersey Audubon Society
794 Rancocas Road
Mount Holly, NJ 08060
(609) 261-2495

World Series of Birding
Cape May Bird Observatory
Box 3
Cape May Point, NJ 08212
(609) 884-2736
Date: May

New Hampshire

Birdwatch America
816 Elm Street, Suite 234
Manchester, NH 03101
(603) 478-3595

New Mexico

Festival of the Cranes
Socorro Chamber
of Commerce
Box 743
Socorro, NM 87801
(505) 835-0424
(505) 835-7600
Date: November

Gila Bird and Nature Festival
Box 1163
Silver City, NM 88062
(800) 828-8291
Date: April

Oregon

Oregon Shorebird Festival
Cape Arago Audubon Society
Box 381
North Bend, OR 97459
Date: September

John Scharff Migratory Bird
Festival and Art Show
Harney Chamber of Commerce
18 West D Street
Burns, OR 97720
(503) 573-2636
Date: April

Birding Events

Pennsylvania

Pocono Environmental
Education Center
RD 2, Box 1010
Dingmans Ferry, PA 18328
(717) 828-2319
Date: year-round

Texas

Hummer/Bird Celebration
Rockport-Fulton
Chamber of Commerce
Rockport, TX 78382
(800) 242-0071
(800) 826-6441
Date: September

Wisconsin

Birds in Art Annual Exhibit
Leigh Yawkey Woodson
Art Museum
700 North 12th Street
Wausau, WI 54401
(715) 845-7010
Date: winter

Fall Flyway and Spectacle
of the Geese
Fond du Lac Convention
and Visitors Bureau
19 West Scott Street
Fond du Lac, WI 54935
(414) 923-3010
Date: October

CANADA

Ontario

Festival of Hawks
Holiday Beach Migration
Observatory
Holiday Beach, Ontario
Date: autumn

INTERNATIONAL

Bonaire

Bonaire Birdwatching
Olympics and Nature Week
Tourism Corporation Bonaire
2039 Ninth Avenue
Ronkonkoma, NY 11779
Date: September

Israel

Migration Festival
International Birding Center
Box 774
Eilat, Israel 88000

Europe

Euro Bird Week
Dutch Birding
Postbus 75611
1070 AP
Amsterdam, Netherlands

Worldwide

International Migratory
Bird Day
Mary Deinlein
Smithsonian Migratory
Bird Center
c/o National Zoo
Washington, DC 20008
(202) 673-4908

or

Rick Bonney
Cornell Laboratory of
Ornithology
159 Sapsucker Woods Road
Ithaca, NY 14850
(607) 254-2400
Date: second Saturday in May

Chapter 4

ORGANIZATIONS FOR BIRDERS

Perhaps because they tend to have orderly minds, birders have a talent for organization. The venerable Audubon movement, at the state and national level, dates back to before the turn of the twentieth century. As the pressure has mounted on the environment, many other organizations have been formed to help preserve birds and the natural world and to promote the art and science of birdwatching.

THE AUDUBON MOVEMENT

The Audubon movement today is slightly confusing to the uninitiated, because it exists on three levels. The National Audubon Society is a nationwide, nonprofit organization devoted to the preservation of the natural world. National headquarters for the more than 500,000 members is in New York City; there is also a lobbying office in Washington, DC. In addition, the National Audubon Society has regional field offices and numerous affiliated local branches. Traveling birders should contact the regional office for information about local offices in the areas that will be visited. The local branches, which are generally run by a small paid staff or entirely by volunteers, organize regular field trips to local hotspots; visitors are always welcome to come along. The National Audubon Society publishes the monthly magazine *Audubon*, with general articles about natural history, and the bird-oriented *American Birds* five times a year.

NATIONAL AUDUBON SOCIETY OFFICES

NATIONAL OFFICE

700 Broadway
New York, NY 10003
(212) 979-3000

Capital Office

666 Pennsylvania Avenue, SE
Washington, DC 20003
(202) 547-9009

Regional Field Offices

Mid-Atlantic (DC, DE, MD, NJ, PA, VA, WV)
1104 Fernwood Avenue, Suite 300
Camp Hill, PA 17011
(717) 763-4985

Northeast (CT, ME, MA, NH, NY, RI, VT)
1789 Western Avenue
Albany, NY 12203
(518) 869-9731

Southeast (AL, FL, GA, KY, MS, NC, SC, TN)
102 East Fourth Avenue
Tallahassee, FL 32303
(904) 222-2473

Great Lakes (IL, IN, KY, MN, OH, WI)
692 North High Street, Suite 208
Columbus, OH 43215
(614) 224-3303

West Central (AR, IA, KS, MO, NE, ND, OK, SD)
200 Southwind Place
Manhattan, KS 66502
(913) 537-4385

Southwest (LA, NM, TX, Guatemala, Mexico, Panama)
2525 Wallingwood, Suite 1505
Austin, TX 78746
(512) 327-1943

Rocky Mountain (AZ, CO, ID, MT, UT, WY)
4150 Darley, Suite 5
Boulder, CO 80303
(303) 499-0219

Western (CAA, NV, OR, WA, Guam)
555 Audubon Place
Sacramento, CA 95825
(916) 481-5332

Alaska and Hawaii
308 G Street, Suite 217
Anchorage, AK 99501
(907) 276-7034

NATIONAL AUDUBON SOCIETY LOCAL CHAPTERS

Many local chapters of the National Audubon Society are quite active, with an extensive program of natural history field trips and other events. Listed below are some—but by no means all—active local chapters, along with many of the statewide Audubon councils.

Arizona

Arizona Audubon Council
505 Morgan Road
Sedona, AZ 86336

Tucson Audubon Society
200 East University Boulevard, Suite 120
Tucson, AZ 85705
(602) 629-0510
Publication: *The Vermilion Flycatcher*

California

Los Angeles Audubon Society
7377 Santa Monica Boulevard
Los Angeles, CA 90046
Publication: *Western Tanager*
(213) 876-0202

Mount Shasta Area
Audubon Society
Box 530
Mount Shasta, CA 96067
(916) 926-4251
(916) 926-3900
Publication: *Endeavor* newsletter

Connecticut

Audubon Council of Connecticut
RR 1, Box 171
Sharon, CT 06069
(203) 364-0520
Publication: *Audubon Update*

Illinois

Audubon Council of Illinois
c/o Bonnie John, Secretary
824 South Dunton
Arlington Heights, IL 60005
(708) 259-5168

Kentucky

Kentucky Audubon Council
c/o Stan Cotton, Secretary
Box 1273
Henderson, KY 42420

Louisiana

Louisiana Audubon Council
355 Napoleon Street
Baton Rouge, LA 70802
(504) 346-8761

Michigan

Michigan Audobon Society
Box 80527
Lansing, MI 48908
Publication: *Michigan Birds and Natural History*

Missouri

Audubon Society of Missouri
c/o Emily Bever, Secretary
704 Bitterfield
Ballwin, MO 63011
(314) 391-6693
Publication: *The Bluebird*

Montana

Montana Audubon Council
Box 595
Helena, MT 59624
(406) 443-3949

New York

New York City
Audubon Society
71 West 23rd Street, Suite 1430
New York, NY 10010
(212) 691-7483
Publication: *The Urban Audubon*

Ohio

Ohio Audubon Council, Inc.
121 Larchmont Road
Springfield, OH 45503

Oregon

Audubon Society of Portland
5151 NW Cornell Road
Portland, OR 97210
(503) 682-6726
Publications: *Audubon Warbler; The Urban Naturalist*

Vermont

Vermont Audubon Council
c/o Judy Peterson, Secretary
RD 1, Box 334A
Middlebury, VT 05753
(802) 388-0126

INDEPENDENT AUDUBON SOCIETIES

A number of states have independent Audubon societies not affiliated with the National Audubon Society. This confusing situation dates back to the start of the Audubon movement at the turn of the century, when numerous conservation organizations sprang up using the Audubon name. Many eventually coalesced into what is now the National Audubon Society, but some remained unaffiliated as state societies with their own local chapters, magazines, sanctuaries and activities. To complicate matters further, the National Audubon Society has chapters in every state, including those with independent Audubons. Fortunately, peaceful coexistence and frequent cooperation seem to be the rule.

ORGANIZATIONS FOR BIRDERS

Audubon Naturalist Society
of the Central Atlantic States
8940 Jones Mill Road
Chevy Chase, MD 20815
(301) 652-9188

Connecticut Audubon Society
118 Oak Street
Hartford, CT 06106
(203) 527-8737
Publication: *Connecticut Audubon* bulletin

Florida Audubon Society
460 Highway 436, Suite 200
Casselberry, FL 32707
(407) 260-8300
Publication: *The Florida Naturalist*

Hawaii Audubon Society
212 Merchant Street, Suite 320
Honolulu, HI 96813
(808) 528-1432
Publication: *Elepaio*

Illinois Audubon Society
Box 608, Whitethorn Road
Wayne, IL 60184
(708) 584-6290
Publications: *Illinois Audubon; The Cardinal News*

Indiana Audubon Society, Inc.
Mary Gray Bird Sanctuary
RR 6, Box 163
Connersville, IN 47331
(317) 827-0908
Publications: *The Indiana Audubon; The Cardinal*

Maine Audubon Society
Gilsland Farm
Box 6009
Falmouth, ME 04105
(207) 781-2330
Publication: *Habitat: Journal of the Maine Audubon Society*

Massachusetts Audubon
Society, Inc.
South Great Road
Lincoln, MA 01773
(617) 259-9500
Publication: *Sanctuary*

Michigan Audubon Society
6011 West St. Joseph, Suite 403
Box 80527
Lansing, MI 48908
(617) 886-9144
Publication: *Jack-Pine Warbler*

Audubon Society
of New Hampshire
Three Silk Farm Road
Box 528B
Concord, NH 03302
(603) 224-9909
Publications: *NH Audubon; Audubon Alert*

New Jersey Audubon Society
Box 125
790 Ewing Avenue
Franklin Lakes, NJ 07417
(201) 891-1211
Publications: *New Jersey Audubon; Peregrine Observer*

Audubon Society of Rhode
Island
12 Sanderson Road
Smithfield, RI 02917
(401) 231-6444
Publication:
Field Notes of RI Birds

THE AMERICAN BIRDING ASSOCIATION

The American Birding Association is oriented toward birders with a serious interest in listing and in traveling to see birds. The goals of the organization are to promote recreational birding, to contribute to the development of bird identification and population study, and to help foster public appreciation of birds and their vital role in the environment. The ABA is also the semiofficial arbiter of North American listing standards and records. Most importantly, the ABA works to promote ethical birding—see the ABA Code of Ethics in Appendix B. Read it carefully and take it to heart.

The ABA now has some 13,000 members representing every North American state, province, and territory as well as 44 countries worldwide. It publishes the bimonthly journal *Birding*, the lively monthly newsletter *Winging It*, and the quarterly newsletter *A Bird's-Eye View* for young birders.

American Birding Association
Box 6599
Colorado Springs, CO 80934
(800) 835-2473
(719) 634-7736

CORNELL LABORATORY OF ORNITHOLOGY

The Laboratory of Ornithology at Cornell University in Ithaca, New York, was founded in 1955 by two eminent ornithologists, Professor Arthur A. Allen and Professor Peter Paul Kellogg. Dr. Allen was a pioneering bird photographer; Dr. Kellogg was a leader in birdsong recording. The innovative programs begun by these two men established the laboratory as an important center for bird study. Today the laboratory is a nonprofit membership organization dedicated to the study and appreciation of birds.

One of the most interesting aspects of the lab is the Library of Natural Sounds, which has the largest collection of birdsong recordings in the world. The lab sponsors numerous ongoing research projects, including breeding bird atlases and Project FeederWatch. Travel programs, courses, and other lab activities are mentioned elsewhere in this book.

The laboratory publishes an excellent quarterly magazine, *Living Bird*.

Cornell Laboratory of Ornithology
159 Sapsucker Woods Road
Ithaca, NY 14850
(607) 254-2400

NATIONWIDE ORGANIZATIONS

American Backyard
Bird Society
Box 10046
Rockville, MD 20849
(301) 309-1431

American Littoral Society
Sandy Hook
Highlands, NJ 07732
(201) 291-0055
Publications: *Underwater Naturalist; Coastal Reporter*

The American
Ornithologists' Union
c/o Division of Birds
National Museum
of Natural History
Washington, DC 20560
(202) 357-1300
Publications: *The Auk; Ornithological Newsletter*

Association of Field
Ornithologists, Inc.
Elissa M. Landre, Secretary
278 Eliot Street
South Natick, MA 01760
(508) 655-6572
Publications: *Journal of Field Ornithology; Ornithological Newsletter*

BirdLife International
(formerly International
Council for Bird Preservation)
U.S. Office
c/o World Wildlife Fund
1250 24th Street, NW
Washington, DC 20037
(202) 778-9563

Nationwide Organizations

Pan American Section
c/o Betsy Trent Thomas,
Secretary
Smithsonian Institution
National Zoological Park
Washington, DC 20008
(202) 673-4717
Publications: *World Birdwatch;
World Bird Conservation*

The Brooks Bird Club, Inc.
707 Warwood Avenue
Wheeling, WV 26003
Publication: *The Redstart*

The Canvasback Society
Box 101
Gates Mills, OH 44040
(216) 443-2340
Publication: *The Canvasbacker*

Center for the Study of
Tropical Birds
218 Conway
San Antonio, TX 78209
(512) 828-5306

Colorado Bird Observatory
13101 Picadilly Road
Brighton, CO 80601
(303) 659-4348

Cooper Ornithological Society
Department of Biology
University of California
Los Angeles, CA 90024
(213) 740-2777
Publication: *The Condor*

Cornell Laboratory of
Ornithology
159 Sapsucker Woods Road
Ithaca, NY 14850
(607) 254-2473
Publication: *Living Bird*

Defenders of Wildlife
1244 19th Street, NW
Washington, DC 20036
(202) 659-9510
Publication: *Defenders*

Delta Waterfowl Foundation
102 Wilmot Road, Suite 410
Deerfield, IL 60015
(708) 940-7776

Ducks Unlimited, Inc.
One Waterfowl Way
Long Grove, IL 60047
(708) 438-4300
Publication: *Puddler Magazine;
Ducks Unlimited Magazine*

The Eagle Foundation
300 East Hickory Street
Apple River, IL 61001
(815) 594-2259

George Miksch Sutton Avian
Research Center, Inc.
Box 2007
Bartlesville, OK 74005
(918) 336-7778
Publication: *The Sutton*

Hawk Migration Association
of North America
c/o Myriam Moore, Secretary
Box 3482
Lynchburg, VA 24503
(804) 847-7811
Publication: *Hawk Migration*

ORGANIZATIONS FOR BIRDERS

Studies

Hawk Mountain Sanctuary
Association
RD 2, Box 191
Kempton, PA 19529
(215) 756-4468
Publication:
Hawk Mountain News

HawkWatch International, Inc.
Box 660
Salt Lake City, UT 84110
(801) 254-8511

Inland Bird Banding
Association
RD 2, Box 26
Wisner, NE 68791
(402) 529-6679
Publication: *North American Bird Bander*

International Crane
Foundation
E-11376
Shady Lane Road
Baraboo, WI 53913
(608) 356-9462
Publications: *ICF Bugle; Cranes, Cranes, Cranes*

The International Osprey
Foundation, Inc.
Box 250
Sanibel, FL 33957
(813) 472 5218

International Wild Waterfowl
Association
5614 River Styx Road
Medina, OH 44256
Publication: *IWWA Newsletter*

Manomet Bird Observatory
Box 936
Manomet, MA 02345
(508) 224-6521

National Bird-Feeding Society
2218 Crabtree
Northbrook, IL 60065
(708) 272-0135
Publication:
The Bird's-Eye reView

National Flyway Council
c/o Jay Lawson, Chairman
Chief Game Warder
Wyoming Game and Fish
Department
5400 Bishop Boulevard
Cheyenne, WY 82006
(307) 777-7735

National Foundation to Protect
America's Eagle
Box 120206
Nashville, TN 37212
(800) 2-EAGLES
(615) 847-4171
Publication:
American Eagle News

The National Wild Turkey
Federation, Inc.
Wild Turkey Building
Box 530
Edgefield, SC 29824
(803) 637-3106
Publications: *Turkey Call Magazine; The Caller*

Nationwide Organizations

National Wildlife Federation
1400 Sixteenth Street NW
Washington, DC 20036
(202) 797-6800
Publications: *International Wildlife; National Wildlife; Your Big Backyard*

National Wildlife Refuge Association
10824 Fox Hunt Lane
Potomac, MD 20854
Publication: *Blue Goose Flyer*

National Wildlife Rehabilitators Association
Carpenter Nature Center
12805 St. Croix Trail
Hastings, MN 55033
(612) 437-9194
Publication: *Wildlife Rehabilitation*

North American Bluebird Society
Box 6295
Silver Spring, MD 20906
(301) 384-2798
Publication: *Sialia*

North American Butterfly Association
c/o Dr. Jeffrey Glassberg
39 Highland Avenue
Chappaqua, NY 10514
Publication: *American Butterflies*

North American Crane Working Group
2550 North Diers Avenue,
Suite H
Grand Island, NE 68803
(308) 384-4633
Publication: *The Unison Call*

North American Falconers Association
c/o Larry Miller, Secretary
305 Long Avenue
North Aurora, IL 60542
Publication: *Hawk Chalk*

North American Loon Fund
Six Lily Pond Road
Gilford, NH 03246
(603) 528-4711
Publication: *Loon Call Newsletter*

Organization for Tropical Studies
Box 90630
Durham, NC 27708
(919) 684-5774

Pacific Seabird Group
410 Peregrine Drive SE
Olympia, WA 98503
Publication: *Pacific Seabird Group Bulletin*

The Peregrine Fund, Inc.
5666 West Flying Hawk Lane
Boise, ID 83709
(208) 362-3716
Publication: *Peregrine Fund Newsletter*

Roger Tory Peterson Institute
110 Marvin Parkway
Jamestown, NY 14701
(716) 665 BIRD

ORGANIZATIONS FOR BIRDERS

Pheasants Forever, Inc.
Box 75473
St. Paul, MN 55175
(612) 481-7142
Publication: *Pheasants Forever*

Platte River
Whooping Crane Trust
2550 North Diers Avenue,
Suite H
Grand Island, NE 68803
(308) 384-4633

Point Reyes Bird Observatory
4990 Shoreline Highway
Stinson Beach, CA 94970
(415) 868-1221
Publications: *Observer; PRBO Newsletter*

Prairie Grouse
Technical Council
Wildlife Research Center
317 West Prospect
Fort Collins, CO 80526
(303) 484-2836

Purple Martin Conservation
Association
Edinboro University of
Pennsylvania
Edinboro, PA 16444
(814) 734-4420
Publication:
Purple Martin Update

Quail Unlimited, Inc.
Box 10041
Augusta, GA 30903
(803) 637-5731
Publication: *Quail Unlimited*

The Raptor Center
University of Minnesota
1920 Fitch Avenue
St. Paul, MN 55108
Publication:
The Raptor Release

Raptor Education
Foundation, Inc.
21901 East Hampden Avenue
Aurora, CO 80013
(303) 680-8500
Publication: *Talon*

Raptor Research
Foundation, Inc.
c/o Jim Fitzpatrick
Carpenter Nature Center
12805 St. Croix Trail
Hastings, MN 55033
(612) 437-4359
Publications: *Journal of Raptor Research; The Kettle*

RARE Center for Tropical
Bird Conservation
1529 Walnut Street
Philadelphia, PA 19102
(215) 568-0516

The Ruffed Grouse Society
451 McCormick Road
Coraopolis, PA 15108
(412) 262-4044
Publication: *RGS*

Society for the Preservation of
Birds of Prey
Box 66070
Los Angeles, CA 90066
(310) 397-8216
Publication:
The Raptor Report

Nationwide Organizations

Society of Tympanuchus
Cupido Pinnatus, Ltd.
930 Elm Grove Road
Elm Grove, WI 53122
(414) 782-6333
Publication: *Boom*

The Trumpeter Swan Society
3800 County Road 24
Maple Plain, MN 55359
(612) 476-4663
Publications: *Trumpetings;
Trumpeter Swan Society*
newsletter

Western Field Ornithologists
c/o Tucson Audubon Society
300 East University Boulevard,
Suite 120
Tucson, AZ 85705
(602) 629-0510

Western Bird Banding
Association
3975 North Pontatoc
Tucson, AZ 85718
(602) 299-1287
Publication: *North American
Bird Bander*

Whooping Crane Conservation
Association, Inc.
1007 Carmel Avenue
Lafayette, LA 70501
(318) 234-6339
Publication: *Grus Americana*

Wild Bird Feeding Institute
1441 Shermer Road
Northbrook, IL 60062
(312) 272-0135

Wildfowl Foundation, Inc.
1101 14th Street, NW, Suite 725
Washington, DC 20005
(202) 371-1808

The Wildfowl Trust of North
America, Inc.
Box 519
Grasonville, MD 21638
(410) 827-6694

Wilson Ornithological Society
c/o Richard C. Banks, President
US Fish and Wildlife Service
National Museum of
Natural History
Washington, DC 20560
(202) 357-1970
Publication:
The Wilson Bulletin

World Bird Sanctuary
Box 270270
St. Louis, MO 63127
(314) 938-6193
Publication: *Mews News*

World Pheasant Association
of U.S.A., Inc.
15545 Regaldo Street
Hacienda Heights, CA 91745
(602) 455-5522
Publication: *WPA Newsletter*
annual journal

The Xerces Society
Ten SW Ash Street
Portland, OR 97204
(503) 222-2788

STATE ORGANIZATIONS

Almost every state has a birding association and local bird clubs in addition to the state and local chapters of the National Audubon Society. These organizations generally have local chapters throughout the state. The parent organization and the local groups sponsor meetings, talks, slide shows, newsletters, field trips, bird walks, and the like.

California Waterfowl
Association
4630 Northgate Boulevard,
Suite 150
Sacramento, CA 95834
(800) 927-3825
(916) 648-1406
Publications: *California Waterfowl; CWAck Tracks*

Carolina Bird Club, Inc.
Box 27647
Raleigh, NC 27611
Publication: *The Chat*

The Connecticut
Ornithological Association
314 Unquowa Road
Fairfield, CT 06430
Publication: *The Connecticut Warbler*

Delmarva Ornithological
Society
Box 4247
Greenville, DE 19807
Publications: *Delmarva Ornithologist; DOS Flyer*

Denver Field Ornithologists
Denver Museum of
Natural History
Denver, CO 80205
(303) 322-7009

Federation of New York State
Bird Clubs, Inc.
c/o Myrna Hemmerick,
Membership Chair
Box 2203
Setauket, NY 11733
Publications: *The Kingbird; New York Birders* newsletter

Georgia Ornithological Society
Box 1684
Cartersville, GA 30120

Iowa Ornithologists' Union
c/o Pam Allen
1601 Pleasant Street
West Des Moines, IA 50265
Publications: *Iowa Bird Life; Newsletter*

Kansas Ornithological Society
c/o Diane Seltman,
Membership Secretary
RR 1, Box 36
Nekoma, KS 67559
Publications: *K.O.S. Bulletin; K.O.S. Newsletter*

Louisiana Ornithological
Society
c/o Mrs. Sammie Rudden
111 Lincoln Road
Monroe, LA 71203

State Organizations

Maryland Ornithological
Society, Inc.
Cylburn Mansion
4915 Greenspring Avenue
Baltimore, MD 21209
(410) 377-8462
Publications: *Maryland Birdlife; The Maryland Yellowthroat*

Minnesota Ornithologists' Union
James Ford Bell Museum of Natural History
Ten Church Street SE
University of Minnesota
Minneapolis, MN 55455
(612) 624-7083
Publications: *The Loon; MOU Newsletter*

Nebraska Ornithologists' Union, Inc.
University of Nebraska
State Museum
W436 Nebraska Hall
Lincoln, NE 68588
(402) 472-6606
Publication: *The Nebraska Bird Review*

New England Hawk Watch
Box 212
Portland, CT 06480

Niagara Peninsula Hawkwatch
c/o John Stevens,
Membership Director
1365 Bayview Avenue, #3
Toronto, Ontario M4G 3A5

Oklahoma Ornithological Society
c/o Aline Romero, Secretary
3730 South Yale Avenue
Tulsa, OK 74135
(918) 742-8366
Publications: *The Scissortail; The Bulletin of the Oklahoma Ornithological Society*

Oregon Field Ornithologists
Box 10373
Eugene, OR 97440

South Dakota Ornithologists' Union
Route 4, Box 252
Brookings, SD 57006
Publications: *South Dakota Bird Notes; Lark Bunting Newsletter*

Tennessee Ornithological Society
Box 402
Norris, TN 37828

Virginia Society of Ornithology
7495 Little River Turnpike,
Suite 201
Annandale, VA 22003
(703) 308-2285
(703) 256-8275
Publications: *The Raven; VSO Newsletter*

Wisconsin Waterfowl
Association, Inc.
Box 792
Waukesha, WI 53187
(414) 524-8460
Publication: *WWA Newsletter*

ORGANIZATIONS FOR BIRDERS

The Wisconsin Society for
Ornithology, Inc.
c/o Carl G. Hayssen, Secretary
6855 North Highway 83
Hartland, WI 53029
(414) 966-2839
Publications: *Badger Birder;
Passenger Pigeon*

CANADIAN ORGANIZATIONS

BirdLife International
(formerly International Council
for Bird Preservation)
Canadian National Section
c/o Canadian Nature
Federation
453 Sussex Drive
Ottawa, Ontario K1N 6Z4
(613) 238-6154

Canadian Nature Federation
453 Sussex Drive
Ottawa, Ontario K1N 6Z4
(613) 238-6154
Publications: *Nature Canada
Magazine; Nature Alert*

Ducks Unlimited Canada
1190 Waverley Street
Winnipeg, Manitoba R3T 2E2
(204) 477-1760

The Nature Conservancy
of Canada
794A Broadview Avenue
Toronto, Ontario M4K 2P7
(416) 469-1701
Publication: *The Ark*
newsletter

Alberta

Calgary Field
Naturalists' Society
1017 19th Avenue NW
Calgary, Alberta T2M 0Z8
Publications: *California
Waterfowl; CWAck Tracks*

Ducks Unlimited Canada
10335 172nd Street
Edmonton, Alberta T5S 1K9
(403) 453-8629

Federation of Alberta
Naturalists
Box 1472
Edmonton, Alberta T5J 2N5
(403) 453-8629
Publication: *Alberta Naturalist*

British Columbia

British Columbia Field
Ornithologists
Box 1018
Surrey, British Columbia
V3S 4P5

The British Columbia
Waterfowl Society
5191 Robertson Road
Delta, British Columbia
V4K 3N2
(604) 946-6980
Publication: *Marsh Notes*

Ducks Unlimited Canada
British Columbia Operation
954A Laval Crescent
Kamloops, British Columbia

Canadian Organizations

V2C 5P5
(604) 374-8307
Publications: *Ducks Unlimited; Conservator*

Federation of British Columbia
Naturalists
1200 Hornby Street
Vancouver, British Columbia
V6Z 2E6
(604) 687-3333

Manitoba

Ducks Unlimited Canada
Manitoba Operation
Box 1160
Stonewall, Manitoba R0C 2Z0
(204) 269-6960
Publication: *The Conservator*

Manitoba Naturalists Society
302-128 James Avenue
Winnipeg, Manitoba R3B ON8
(204) 943-9029

New Brunswick

New Brunswick Federation
of Naturalists
c/o New Brunswick Museum
277 Douglas Avenue
Saint John, New Brunswick
(506) 693-1196
E2K 1E5

Newfoundland and Labrador

Newfoundland Natural
History Society
Box 1013
St. John's, Newfoundland
A1C 5M3

Nova Scotia

Ducks Unlimited Canada
Eastern Canada Operation
Nine Havelock Street
Amherst, Nova Scotia B4H 3Z5
(902) 667-8726

Nova Scotia Bird Society
Nova Scotia Museum
1747 Summer Street
Halifax, Nova Scotia B3H 3A6
(902) 429-4610

Ontario

Ducks Unlimited Canada
240 Bayview Drive, Unit 240
Barrie, Ontario L4N 4Y8
(705) 726-3825

Federation of Ontario
Naturalists
355 Lesmill Road
Don Mills, Ontario M3B 2W8
(416) 444-8419
Publication: *Seasons*

Jack Miner Migratory Bird
Foundation, Inc.
Box 39
Kingsville, Ontario N9Y 2E8
(519) 733-4034

Prince Edward Island

Natural History Society of
Prince Edward Island
Box 2346
Charlottestown, Prince Edward
Island C1A 1R4

Quebec

Ducks Unlimited Canada
Quebec Operation
710 Bouvier Street, Suite 260
Quebec, Quebec G2J 1A7

Province of Quebec Society for
the Protection of Birds, Inc.
4832 de Maisonneuve
Boulevard West
Montreal, Quebec H3Z 1M5
(514) 937-0224
Publications: *Tchebec*;
newsletter

Saskatchewan

Ducks Unlimited Canada
Saskatchewan Operation
Box 4465
1606 Fourth Avenue
Regina, Saskatchewan
S4P 3W7
(306) 569-0424

Saskatchewan Natural
History Society
Box 414
Raymore, Saskatchewan
S0A 3J0
(306) 746-4544

Yukon Territory

Yukon Bird Club
c/o Cameron Eckert
14 11th Avenue
Whitehorse, Yukon YIA 4H6

Yukon Conservation Society
T.C. Richards Building
Box 4163
Whitehorse, Yukon Territory
Y1A 2C6

INTERNATIONAL ORGANIZATIONS

The African Bird Club
BirdLife International
Wellbrook Court
Girton Road
Cambridge CB3 0NA
England

BirdLife International
(formerly International Council
for Bird Preservation)
32 Cambridge Road
Girton, Cambridge CB3 0PJ
England
0223 277318

The British
Ornithologists Union
c/o The Natural
History Museum
Sub-Department of
Ornithology
Tring, Herts HP23 6AP
England
0442 890080

British Trust for Ornithology
The Nannery
Nannery Place
Thetford, Norfolk IP24 2P4
England
0842 750050

International Organizations

Danish Bird Association
(Dansk Ornitologisk)
Vesterbrogade 140
DK-1620 Copenhagen V
Denmark
45-31-31-4404

East African Wildlife Society
Box 20110
Nairobi, Kenya
(254) 2-227047

The Hawk and Owl Trust
c/o Birds of Prey Section
London Zoo
Regent's Park, London
NW1 4RY
England
71 722 3333

Irish Wildbird Conservancy
Rutledge House,
8 Longford Place
Monkstown, County Dublin
Ireland
(01) 280-4322

Neotropical Bird Club
c/o Rob Williams, Publicity
Officer
The Lodge
Sandy, Bedfordshire SG19 2DL
England
0767 680 551

Oriental Bird Club
c/o Robert S. Kennedy, Ph.D.
Cincinnati Museum of
Natural History
1720 Gilbert Avenue
Cincinnati, OH 45202
(513) 287-7000

Partners in Flight—Aves
de las Americas
c/o Peter Stangel
National Fish and
Wildlife Foundation
18th and C Streets NW,
Suite 2556
Washington, DC 20240
(202) 857-0166

Polish Society for the
Protection of Birds
c/o BirdLife International
32 Cambridge Road
Girton, Cambridge CB3 0PJ
England
0223 277318

Royal Society for the
Protection of Birds
The Lodge
Sandy, Bedfordshire SG12 2OL
England
0767 680551

Swedish Bird Association
(Sveriges Ornitologiska
Forening)
Skeppargatan 19
Box 14219
S-104 40 Stockholm
Sweden
86-62-6434

Wild Bird Society of Taipei
6, Alley 13, Lane 295
Fu-Shin South Road, Section 1
Taipei, Taiwan, R.O.C
886-2-325-9190

Chapter 5

OPTICS FOR BIRDERS

Investing in a really good pair of binoculars and an excellent spotting scope pays amazing dividends in birding pleasure. These essential items cost significant amounts of money, however, and the decisions about what to get and where to purchase it should not be made lightly. As always with any major purchase, be informed and shop around.

To get information about a specific product, write directly to the manufacturer. The company will respond with large quantities of brochures, fact sheets, reviews, and other information designed to prove that its products are clearly superior to all others. The final choice of optics often depends, however, on fairly subjective judgments based on the "feel" of the product—how well the binoculars fit your hands, for example. The only way to find this out is to physically try the products. This can be done at a large optics retailer if you happen to live near one. If you don't, try looking through the binoculars and scopes of everybody in your birding group.

The widest selection of in-stock items is generally found at specialty optics dealers, many of whom are oriented specifically toward birders. These dealers usually have retail stores, but the bulk of their business is through the mail. They offer discounted prices, quick shipping, and good advice. Write for catalogs.

After you've bought your new binoculars, don't relegate the old pair to the back of a cupboard. These binoculars can have a valuable new life helping citizens in Latin America learn about their wildlife. To donate binoculars in any condition for conservation education in Latin America, contact:

Optics Manufacturers

Gary Filerman *or*
1322 Banquo Court
McLean, VA 22102
(703) 356-9033

Manomet Bird *or*
Observatory
Birder's Exchange
Box 936
Manomet, MA 02345
(508) 224-6521

George Shillinger
Birdlife International
c/o World Wildlife
Fund
1250 24th Street NW
Washington, DC
20037
(202) 778-9563

For field tests, technical information, and other data about birding optics, subscribe to the quarterly newsletter:

Better View Desired
Whole Life Systems
Box 162
Rehoboth, NM 87322
(505) 863-4751

Listed below are manufacturers of binoculars, spotting scopes, and optical accessories. Also listed are specialty optics dealers oriented toward birders. Because these dealers are just a phone call away and will ship anywhere, they are listed alphabetically and not by state.

Optics Manufacturers

Bausch & Lomb
9200 Cody
Overland Park, KS 66214
(800) 423-3537
Products: Binoculars (Elite, Custom, NatureView); spotting scopes (Elite)

Brunton U.S.A.
Riverton, WY 82501
Products: Binoculars (Eterna)
(307) 856-6559

Celestron International
2835 Columbia Street
Torrance, CA 90503
(310) 328-9560
Products: Binoculars and spotting scopes

Fujinon Inc.
Ten High Point Drive
Wayne, NJ 07470
(201) 633-5600
Products: Binoculars

aus Jena
Europtik, Ltd.
Box 319
Dunmore, PA 18512
(800) 347-6049
(717) 347-6049
Products: Binoculars and spotting scopes

OPTICS FOR BIRDERS

Kowa Optimed Inc.
20001 South Vermont Avenue
Torrance, CA 90502
(213) 327-1913
Products: Spotting scopes
(50 mm, 60 mm, 77 mm)

Leica Camera Inc.
156 Ludlow Avenue
Northvale, NJ 07647
(800) 222-0118
(201) 767-7500
Products: Binoculars (Trinovid, Ultra)

Minolta Corporation
101 Williams Drive
Ramsey, NJ 07446
(201) 825-4000
Products: Binoculars
(Standard)

Mirador Optical Corporation
Box 11614
Marina Del Rey, CA 90295
Products: Binoculars (A Series)

Nikon Sport Optics
1300 Walt Whitman Road
Melville, NY 11747
(800) 247-3464
Products: Binoculars, spotting scopes

Optolyth U.S.A., Inc.
18805 Melvista Lane
Hillsboro, OR 97123
(800) 447-6881
Products: Spotting scopes

Pentax Corporation
35 Inverness Drive East
Englewood, CO 80112
(303) 799-8000
Products: Binoculars

Quester Corporation
Box 59
New Hope, PA 18938
(215) 862-5277

Redfield
5800 East Jewell Avenue
Denver, CO 80224
(303) 757-6411
Products: Binoculars and spotting scopes

Swarovski Optik
One Wholesale Way
Cranston, RI 02920
(800) 426-3089
Products: Binoculars and spotting scopes

Swift Instruments Inc.
952 Dorchester Avenue
Boston, MA 02125
(617) 436-2960
Products: Binoculars
(Audubon, Compact Audubon, Ultra-Lite)

Carl Zeiss Optical, Inc.
1015 Commerce Street
Petersburg, VA 23803
(804) 861-0033
Products: Binoculars (Dialyt)

Optics Retailers

American Birding
Association Sales
Box 6599
Colorado Springs, CO 80934
(800) 634-7736

Berger Bros.
209 Broadway
Amityville, NY 11701
(516) 264-4160

Birding
Box 4405
Halfmoon, NY 12065
(518) 664-2011

Camera Corner of Iowa
(800) 762-4282
(319) 391-4778

Camera One
1918 Robinhood
Sarasota, FL 34231
(800) 995-1302
(813) 924-1302

Christophers, Ltd.
2401 Tee Circle, Suite 105
Norman, OK 73069
(800) 356-6603
(405) 364-4898

City Camera
15336 West Warren
Dearborn, MI 48128
(800) 359-5085

Eagle Optics
716 South Whitney Way
Madison, WI 53719
(800) 289-1132
(608) 271-4751

Executive Photo & Electronics
120 West 31 Street
New York, NY 10001
(800) 223-7323
(212) 947-5290

Focus Camera and Video
4419-21 13th Avenue
Brooklyn, NY 11219
(800) 221-0828
(718) 436-6262

Robert Manns and Associates
Atlanta, GA 30318
(404) 350-9791

F.C. Meichsner Co., Inc.
182 Lincoln Street
Boston, MA 02111
(800) 321-VIEW

Mirakel Optical Co., Inc.
331 Mansion Street
West Coxsackie, NY 12192
(518) 731-2610

National Camera Exchange
9300 Olson Memorial Highway
Golden Valley, MN 55427
(800) 624-8107
(612) 546-6831

Nelson-Halik Optics
2087 Nobleshire Road
Columbus, OH 43229
(614) 882-7006

Orion Binoculars and
Telescopes
2450 17th Avenue
Santa Cruz, CA 95061
(800) 447-1001

Phil's Camera Service
11517 South Dixie Highway
Miami, FL 33156
(305) 238-7842

Scope City
Box 440
Simi Valley, CA 93065
(800) 235-3344
(805) 522-6701

S.A. Wentling Co.
Box 355D
Hershey, PA 17033
(717) 533-2468

Wholesale Optics of
Pennsylvania
RR 6, Box 6329
Moscow, PA 18444
(717) 842-1500

Optics Repair

F.C. Meichsner Co., Inc.
182 Lincoln Street
Boston, MA 02111
(800) 321-VIEW

Mirakel Optical Co., Inc.
331 Mansion Street
West Coxsackie, NY 12192
(518) 731-2610

Redlich Binocular and Optical
Repair Service
711 West Broad Street
Falls Church, VA 22046
(703) 241-4077

Optics Accessories

Ballard Industries
10271 Lockwood Drive, Suite B
Cupertino, CA 95014
(800) 852-2468 (fax)
Products: Lens hoods

The Birder's Connection
2521 College Road
Downers Grove, IL 60516
(708) 852-9615
Products: The Link
binoculars strap

Bogen Photo Corp.
565 East Crescent Avenue
Ramsey, NJ 07446
(201) 818-9500
Products: Bogen tripods and
tripod heads

Marlab Specialties
Box 4019
San Rafael, CA 94903
(800) 237-4293
Product: ProHarness
binoculars strap

Op/Tech USA
290 Arden Drive
Belgrade, MT 59714
(800) 541-4916
(406) 388-1377
Products: Carrying straps,
binoculars caps

OptiSoc
Box 422
Youngsville, NC 27596
Product: Field covering
for binoculars

Bird Sound Amplification Devices

Leonard Rue Enterprises
138 Millbrook Road
Blairstown, NJ 07825
(908) 362-6616
Products: Catalog of items for birders and bird photographers

Visual Departures, Ltd.
1641 Third Avenue
New York, NY 10128
(800) 628-2003
Products: Steadybag "beanbag" support for optics

Tenth House Gear
RR 1
Golden Lake, Ontario K0J 1X0
(613) 625-2263
Products: Grebe scope/tripod carrying case

BIRD SOUND AMPLIFICATION DEVICES

Amplification devices are designed to bring in the sounds of birds, making them easier to locate and identify.

Manufacturers

BPA Marketing, Inc.
3519 Bigelow Boulevard
Pittsburgh, PA 15213
(800) 221-1196
Product: Songscope 2 microphone/amplifier

Natural Technology Industries
Box 582
Youngstown, OH 44501
(216) 742-6206
Products: Bird Bug wildlife monitor and amplifier

NatureSound Research
Box 84
Ithaca, NY 14851
Product: SongFinder birdsong listening device

Sonic Technology Products, Inc.
120 Richardson Street
Grass Valley, CA 95945
(800) 247-5548
Product: SuperEar Sound Enhancer

Walker's "Game Ear" Inc.
PO Box 1069
Media, PA 19063
Product: Game Ear sound amplification device

Chapter 6

THE EDUCATED BIRDER

Birders always want to know more. In addition to formal courses, seminars, photo workshops, and the like, birders can educate themselves through visits to museums and zoos. This chapter includes information on all.

COURSES AND OTHER PROGRAMS

The opportunities for ongoing education in ornithology and related subjects are many, as listed below. In addition to these sources, check with local colleges, community colleges, bird clubs, and environmental organizations. Many offer courses on birds and related subjects.

HOME STUDY

The well-known Cornell home-study course is an excellent way to learn college-level ornithology at your own pace under the guidance of an expert education staff. For more information, write to:

Cornell Laboratory of Ornithology
Home Study Course
159 Sapsucker Woods Road
Ithaca, NY 14850
(607) 254-2444

NATIONAL AUDUBON SOCIETY EDUCATION CENTERS

Richardson Bay Wildlife
Sanctuary and Whittell
Education Center
376 Greenwood Beach Road
Tiburon, CA 94920
(415) 388-2524

National Environmental
Education Center
613 Riversville Road
Greenwich, CT 06831
(203) 869-5272

Northeast Audubon Center
RR 1, Box 171
Sharon, CT 06069
(203) 364-0520

Todd Wildlife Sanctuary
Keene Neck Road
Medomak, ME 04551
(207) 529-5148

Audubon Center of the
Northwoods
Route 1

Sandstone, MN 55072
(612) 245-2648

Randall Davey Audubon
Center
Box 9314
Santa Fe, NM 87504
(505) 983-4609

Theodore Roosevelt Memorial
Bird Sanctuary
134 Cove Road
Oyster Bay, NY 11771
(516) 922-3200

Aullwood Audubon
Center and Farm
1000 Aullwood Road
Dayton, OH 45414
(513) 890-7360

Schlitz Audubon Center
1111 East Brown Deer Road
Milwaukee, WI 53217
(414) 352-2880

NATIONAL AUDUBON SOCIETY INTERN PROGRAM

The National Audubon Society Intern Progam was created to help individual sanctuaries meet their need for additional personnel. In turn, the program is designed to benefit the intern by providing hands-on experience in all phases of wildlife sanctuary work. Internships are available at Borestone Mountain Sanctuary, Maine; Clyde E. Buckley Sanctuary, Kentucky; Corkscrew Swamp Sanctuary, Florida; Francis Beidler Forest Sanctuary, South Carolina; Emily W. Miles Sanctuary, Connecticut; Appleton–Whittell Research

Sanctuary, Arizona; Starr Ranch Sanctuary, California; Maine Coastal Islands Sanctuary, Maine; and Constitution Marsh Sanctuary, New York. For more information about the intern program, contact:

National Audubon Society
Sanctuary Department
93 West Cornwall Road
Sharon, CT 06069
(203) 364-0048

COURSES, SEMINARS, INSTITUTES, AND OTHER PROGRAMS

Arizona

Cochise College
Conference Center
Douglas, AZ 85607
(800) 966-7943, extension 316

Tucson Audubon Society
30A North Tucson Boulevard
Tucson, AZ 85710
Programs: Institute of Desert Ecology, Institute of Marine and Coastal Ecology

Maine

College of the Atlantic
105 Eden Street
Bar Harbor, ME 04609
(207) 288-5015
Program: Sea birds study

Eagle Hill Wildlife Research Station
Steuben, ME 04680
(207) 546-2821
Programs: Field ornithology

Institute for Field Ornithology
University of Maine at Machias
Nine O'Brien Avenue
Machias, ME 04654
(207) 258-3313, extension 289
Programs: Field ornithology

Massachusetts

Bird Nantucket
Box 1182
Nantucket, MA 02554
Program: Bird banding research

Massachusetts Audubon Society
Box 236
South Wellfleet, MA 02663
(508) 349-2615
Programs: Coastal birding, nature photography

Minnesota

Birding Workshops
Extended Education
University of Wisconsin at LaCrosse

Courses, Seminars, Institutes, and other Programs

LaCrosse, WI 54601
(608) 785-8569
Programs: Bird identification

Montana

Big Sky Bird Sounds and
Birding Workshop
c/o Yellowstone Wildlife Tours
Box 546
Yellowstone National Park,
WY 82190
(307) 344-7415
Program: Bird sounds and
identification

New Jersey

Institute for Field Ornithology
University of Maine at Machias
Nine O'Brien Avenue
Machias, ME 04654
(207) 258-3313, extension 289
Programs: Field ornithology,
raptors

Pennsylvania

Pocono Environmental
Education Center
RD 2, Box 1010
Dingmans Ferry, PA 18328
(717) 828-2319
Programs: Hawk watches, field
ornithology

Utah

Canyonlands Field Institute
Box 68
Moab, UT 84532
(801) 259-7750
Programs: Introductory birding

Wyoming

Big Sky Bird Sounds and
Birding Workshop
c/o Yellowstone Wildlife Tours
Box 546
Yellowstone National Park,
WY 82190
(307) 344-7415
Program: Bird sounds and
identification

North America

Birding By Ear
Dick Walton
35 Stacey Circle
Concord, MA 01742
(508) 369-3729
Program: Bird sounds

Elderhostel
75 Federal Street
Boston, MA 02110
Program: Bird identification for
people over 60

VOLUNTEER OPPORTUNITIES

One the very best ways to learn about birds is by participating as a volunteer in a research program. You don't necessarily have to know a lot about birds to be helpful—all you need is interest and enthusiasm. For an annual listing of ornithological volunteer opportunities, contact:

Winging It
American Birding Association
Box 6599
Colorado Springs, CO 80934
(800) 634-7736

SUMMER YOUTH PROGRAMS

Summer birdwatching camps for boys and girls ages 13 to 17 are offered by Victor Emanuel Nature Tours (VENT) in Arizona and Mexico. For more information, contact:

Birding Camps
Victor Emanuel Nature Tours (VENT)
Box 33008
Austin, TX 78764

For information on American Birding Association scholarships to attend the VENT camps, contact:

Judith Toups
Chair, ABA Education Committee
Four Hartford Place
Gulfport, MS 39507

Summer programs in field ornithology for 14- to 16-year-olds are offered by:

Colorado Bird Observatory
13101 Picadilly Road
Brighton, CO 80601

Summer programs for high-school students with an interest in ecological and environmental studies are offered by:

Maine Island Ecology
Academy of Natural Sciences
1900 Benjamin Franklin Parkway
Philadelphia, PA 19103
(215) 299-1000

BIRD PHOTOGRAPHY WORKSHOPS

Many outstanding tours, workshops, and seminars for nature photographers are offered by a multitude of experienced leaders—far too many to list here. (Check the advertisements in the birding and nature-photography magazines.) Workshops oriented specifically to bird photography are somewhat less common. Some of those offered on a regular basis at various times and places are listed below. Contact the workshop leader for details.

Allaman's Montana
Photographic Adventures
West Fork Road
Darby, MT 59829
(406) 821-3763

Mike Blair
502 Haskell Street
Pratt, KS 67124
(316) 672-2621

Patricia Caulfield
115 West 86th Street
New York, NY 10024
(212) 362-1951

Carl Kurtz
1562 Binford Avenue
St. Anthony, IA 50239
(515) 477-8364

Peter La Tourrette
1019 Loma Prieta Court
Los Altos, CA 94024
(415) 961-2741

Arthur Morris/Birds as Art
1455 Whitewood Drive
Deltona, FL 32725
(407) 860-2013

Osprey Photo Workshops
Irene Hinke Sacilotto
2719 Berwick Avenue
Baltimore, MD 21234
(410) 426-5071

Jeff Rich Photography
Workshops
18962 River Ranch Road
Anderson, CA 96007
(916) 241-6153

John Shaw
5504 Rutledge Drive
Greensboro, NC 27455
(919) 282-2329

Summit Photographic
Workshops
Box 24571
San Jose, CA 95154
(408) 265-4627

Joseph Van Os Photo Safaris
Box 655
Vashon, WA 98070
(206) 463-5383

Larry West
24 West Barnes Road
Mason, WI 48854
(517) 676-1890

ZOOS AND AVIARIES

Captive birds don't count on a life list, but they are still extremely interesting, since observations of their appearance and behavior can help with field identification later. Zoos are also sometimes the only place you can ever see an endangered bird such as the California condor.

Almost all zoos and aquariums have some bird species on display. Indeed, as of 1992 the 150 members of the American Association of Zoological Parks and Aquariums collectively have 12,216 bird species and 53,099 individual specimens. The list below is somewhat arbitrarily limited to those zoos with more than 100 bird species. The only exception is the Arizona Sonora Desert Museum, whose walk-in hummingbird aviary is a rare experience.

Zoo hours, exhibits, and activities tend to vary with the season; call in advance when planning a visit.

UNITED STATES

Arizona

Arizona Sonora
Desert Museum
2021 North Kinney Road
Tucson, AZ 85743
(602) 883-1380
Birds: 68 species, 284 specimens

The Phoenix Zoo
105 East Campo Allegro
Tempe, AZ 85281
(602) 273-1341
Birds: 145 species, 604 specimens

Wildlife World Zoo
16501 West Northern Avenue
Litchfield Park, AZ 85340
(602) 935-WILD
Birds: 161 species, 899 specimens

California

Los Angeles Zoo
5333 Zoo Drive
Los Angeles, CA 90027
(213) 666-4650
Birds: 261 species, 739 specimens

San Diego Wild Animal Park
15500 San Pasqual Valley Road
Escondido, CA 92027
(619) 747-8702
Birds: 315 species, 1,517 specimens

San Diego Zoo
Zoo Place and Park Boulevard
San Diego, CA 92103
(619) 231-1515
Birds: 485 species, 1,992 specimens

Zoos and Aviaries

San Francisco Zoological
Gardens
One Zoo Road
San Francisco, CA 94132
(415) 753-7080
Birds: 150 species, 405
specimens

Sea World of California
1720 South Shores Road
San Diego, CA 92109
(619) 222-6363
Birds: 133 species, 1,662
specimens

Colorado

Denver Zoological Gardens
City Park
Denver, CO 80205
(303) 331-4100
Birds: 172 species, 708
specimens

District of Columbia

National Zoological Park
3000 Block of Connecticut
Avenue NW
Washington, DC 20008
(202) 673-4721
Birds: 172 species, 845
specimens

Florida

Busch Gardens
3000 Busch Boulevard
Tampa, FL 33612
(813) 988-5171
Birds: 226 species, 1,806
specimens

Discovery Island
Zoological Park
Lake Buena Vista, FL 32830
(407) 824-2875
Birds: 100 species, 472
specimens

Lowry Park Zoological Garden
7530 North Boulevard
Tampa, FL 33604
(813) 935-8552
Birds: 102 species, 335
specimens

Miami Metrozoo
12400 SW 152nd Street
Miami, FL 33177
(305) 251-0401
Birds: 149 species, 662
specimens

Parrot Jungle and Gardens
11000 SW 57th Avenue
Miami, FL 33156
(305) 666-7834
Birds: 101 species, 1,000
specimens

Sea World of Florida
7007 Sea World Drive
Orlando, FL 32821
(407) 351-3600
Birds: 104 species, 1,300
specimens

The Zoo
5701 Gulf Breeze Parkway
Gulf Breeze, FL 32561
(904) 932-2229
Birds: 112 species, 400
specimens

Hawaii

Honolulu Zoo
151 Kapahulu Avenue
Honolulu, HI 96815
(808) 971-7175
Birds: 102 species, 422 specimens

Illinois

Chicago Zoological Park
(Brookfield Zoo)
3300 Golf Road
Brookfield, IL 60513
(708) 485-0263
Birds: 123 species, 485 specimens

Lincoln Park Zoological Gardens
2200 North Cannon Drive
Chicago, IL 60614
(312) 294-4662
Birds: 136 species, 587 specimens

Indiana

Mesker Park Zoo
Bement Avenue
Evansville, IN 47712
(812) 428-0715
Birds: 103 species, 380 specimens

Kansas

Sedgwick County Zoo and Botanical Garden
5555 Zoo Boulevard
Wichita, KS 67212
(316) 942-2213
Birds: 129 species, 566 specimens

Louisiana

Audubon Park and Zoological Garden
6500 Magazine Street
New Orleans, LA 70118
(504) 861-2537
Birds: 139 species, 531 specimens

Greater Baton Rouge Zoo
Greenwood Park, Highway 19
Baker, LA 70704
(504) 775-3877
Birds: 123 species, 496 specimens

Maryland

Baltimore Zoo
Druid Hill Park,
Mansion House
Baltimore, MD 21217
(301) 396-7102
Birds: 103 species, 617 specimens

Minnesota

Minnesota Zoological Garden
13000 Zoo Boulevard
Apple Valley, MN 55124
(612) 431-9200
Birds: 122 species, 659 specimens

Missouri

St. Louis Zoological Park
Forest Park
St. Louis, MO 63110
(314) 781-0900
Birds: 227 species, 967 specimens

Nebraska

Omaha's Henry Doorly Zoo
3701 South Tenth Street
Omaha, NE 68107
(402) 733-8401
Birds: 193 species, 672 specimens

New Mexico

Rio Grande Zoological Park
903 Tenth Street, SW
Albuquerque, NM 87102
(505) 843-7413
Birds: 153 species, 634 specimens

New York

New York Zoological Park
(Bronx Zoo)
185th Street and Southern Boulevard
Bronx, NY 10460
(718) 220-5100
Birds: 320 species, 1,130 specimens

Ohio

Cincinnati Zoo and Botanical Garden
3400 Vine Street
Cincinnati, OH 45220
(513) 281-4701
Birds: 196 species, 890 specimens

Cleveland Metroparks Zoological Park
3900 Brookside Park Drive
Cleveland, OH 44109
(216) 661-6500
Birds: 208 species, 837 specimens

Sea World of Ohio
1100 Sea World Drive
Aurora, OH 44202
(216) 562-8101
Birds: 100 species, 525 specimens

Toledo Zoological Gardens
2700 Broadway
Toledo, OH 43609
(419) 385-5721
Birds: 120 species, 282 specimens

Oklahoma

Oklahoma City Zoological Park
2101 NE 50th Street
Oklahoma City, OK 73111
(405) 424-3344
Birds: 143 species, 385 specimens

Pennsylvania

National Aviary (formerly Pittsburgh Aviary)
Allegheny Commons West

THE EDUCATED BIRDER

Pittsburgh, PA 15212
(412) 323-7235
Birds: 245 species, 750 specimens

Philadelphia Zoological Garden
34th Street and Girard Avenue
Philadelphia, PA 19104
(215) 243-1100
Birds: 201 species, 727 specimens

South Carolina

Riverbanks Zoological Park
500 Wildlife Parkway
Columbia, SC 29210
(803) 779-8717
Birds: 128 species, 473 specimens

Tennessee

Memphis Zoological Garden and Aquarium
2000 Galloway Avenue
Memphis, TN 38112
(901) 726-4787
Birds: 110 species, 423 specimens

Texas

Dallas Zoo
621 East Clarendon Drive
Dallas, TX 75203
(214) 670-6825
Birds: 141 species, 523 specimens

Fort Worth Zoological Park
2727 Zoological Park Drive
Fort Worth, TX 76110
(817) 870-7050
Birds: 120 species, 545 specimens

Houston Zoological Gardens
1513 North MacGregor
Houston, TX 77030
(713) 525-3300
Birds: 202 species, 912 specimens

Gladys Porter Zoo
500 Ringgold Street
Brownsville, TX 78529
(512) 546-7187
Birds: 160 species, 527 specimens

San Antonio Zoological Gardens and Aquarium
3903 North St. Mary's Street
San Antonio, TX 78212
(512) 734-7184
Birds: 224 species, 1,190 specimens

Utah

Hogle Zoological Garden
2600 Sunnyside Avenue
Salt Lake City, UT 84108
(801) 582-1632
Birds: 139 species, 571 specimens

Washington

Woodland Park Zoological Gardens

5500 Phinney Avenue North
Seattle, WA 98103
(206) 684-4880
Birds: 111 species, 335 specimens

CANADA

Alberta

Calgary Zoo, Botanical Garden & Prehistoric Park
Box 3036, Station B
Calgary, Alberta T2M 4R8
(403) 232-9300
Birds: 147 species, 533 specimens

Ontario

Metropolitan Toronto Zoo
West Hill
Toronto, Ontario M1E 4R5
(416) 392-5900
Birds: 144 species, 581 specimens

MUSEUMS WITH SIGNIFICANT ORNITHOLOGICAL COLLECTIONS

In addition to exhibits detailing the natural history of birds, many science museums have extensive collections of specimen birds, eggs, and other items of interest to serious birders.

Museums also often sponsor ongoing avian research projects. Access to the specimen collections may be restricted to researchers, but it is sometimes possible for interested individuals or birding groups to arrange a behind-the-scenes visit.

The museums listed below all have at least 100,000 research specimens. The American Museum of Natural History in New York City has some 900,000 specimens; the Field Museum in Chicago has about 300,000.

When visiting the museums, call in advance to determine hours. Many museums request a voluntary contribution. If possible, give at least the suggested amount.

California

California Academy of Sciences
Golden Gate Park
San Francisco, CA 94118
(415) 221-5100

Natural History Museum of Los Angeles County
900 Exposition Boulevard
Los Angeles, CA 90007
(213) 744-3468

San Bernardino County Museum
2024 Orange Tree Lane
Redlands, CA 92373
(714) 798-8570

San Diego Natural
History Museum
1788 El Prado
Balboa Park
San Diego, CA 92101
(619) 232-3821

Colorado

Denver Museum
of Natural History
2001 Colorado Boulevard
Denver, CO 80205
(303) 370-6357

Connecticut

Peabody Museum
of Natural History
Yale University
170 Whitney Avenue
New Haven, CT 06511
(203) 432-3750

Delaware

Delaware Museum of
Natural History
4840 Kennett Pike
Wilmington, DE 19807
(302) 658-9111

District of Columbia

National Museum of
Natural History
10th Street and Constitution
Avenue NW
Washington, DC 20560
(202) 357-1300

Florida

Florida Museum of
Natural History
University of Florida
Gainesville, FL 32611
(904) 392-1721

Hawaii

Bishop Museum
1525 Bernice Street
Honolulu, HI 96817
(808) 847-3511

Illinois

Field Museum of
Natural History
Roosevelt Road and Lake
Shore Drive
Chicago, IL 60605
(312) 922-9410

Louisiana

Museum of Natural Science
Louisiana State University
119 Foster Hall
Baton Rouge, LA 70803
(504) 388-2855

Michigan

University of Michigan
Museum of Zoology
1109 Geddes
Ann Arbor, MI 48109
(313) 764-0476

Museums with Significant Ornithological Collections

New York

American Museum of
Natural History
Central Park West
at 79th Street
New York, NY 10024
(212) 769-5000

Pennsylvania

The Academy of Natural
Sciences of Philadelphia
1900 and Ben Franklin
Parkway
Philadelphia, PA 19103
(215) 299-1000

Carnegie Museum
of Natural History
4400 Forbes Avenue
Pittsburgh, PA 15213
(412) 622-3131

Chapter 7

BOOKS, SOFTWARE, AND BEYOND

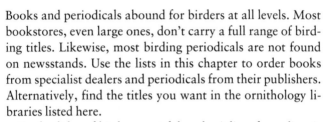

Books and periodicals abound for birders at all levels. Most bookstores, even large ones, don't carry a full range of birding titles. Likewise, most birding periodicals are not found on newsstands. Use the lists in this chapter to order books from specialist dealers and periodicals from their publishers. Alternatively, find the titles you want in the ornithology libraries listed here.

Study slides of birds are useful tools. A list of suppliers is found in this chapter.

Software for listing and other purposes is a burgeoning area in birding. A listing of available software closes this chapter.

BOOKS FOR BIRDERS

Any reasonably well-stocked bookstore will carry a few of the most popular American field guides and books about birding, but for anything beyond that a specialist may be needed. Listed below are some reliable mail-order sources of current titles and out-of-print and antiquarian books; birding book clubs are also listed. Many nature stores also carry a good selection of birding books—see the section on nature gear for more information.

Birders often receive books as gifts. While the thought is always appreciated, sometimes the gift duplicates a title already on the shelf or is irrelevant to your birding interests. A good way to clear out your birding library is to donate useful but unwanted volumes to Books for Nature, an organization that helps promote conservation biology in developing countries. Send your extra books to:

> Books for Nature
> c/o Bernie Tershy
> Neurobiology and Behavior
> Seeley G. Mud Hall
> Cornell University
> Ithaca, NY 14853

Booksellers

American Birding
Association Sales
Box 6599
Colorado Springs, CO 80934
(800) 634-7736
Specialties: Current bird books, finding guides

American Wild Bird
Company Showroom
591 Hungerford Drive
Rockville, MD 20850
(301) 279-8999
Specialties: current bird books

Audubon Naturalist
8940 Jones Mills Road
Chevy Chase, MD 20815
(301) 652-9188
Specialties: current bird books

Buteo Books
Route 1, Box 242
Shipman, VA 22971
(800) 722-2460
(804) 263-8671
Specialties: current and out-of-print bird books

BWD Direct
PO Box 110
Marietta, OH 45750
(800) 879-2473
Specialties: current bird books

Centre de Conservation
de la Faune Ailée
7950 Rue de Marseille
Montreal, Quebec H1L 1N7
(514) 351-5496
Specialties: current bird books

The Chickadee Nature Store
1330-L Wirt
Houston, TX 77055
(713) 956-2670
Specialties: current bird books

Noriko Ciochon
1025 Keokuk Street
Iowa City, IA 52240
(319) 354-9088
Specialties: rare and out-of-print bird books

Alan Contereas Books
2254 Crestview Drive South
Salem, OR 97302

(503) 371-3458
Specialties: current and out-of-print bird books

The Crow's Nest Bookshop
Cornell Laboratory of Ornithology
159 Sapsucker Woods Road
Ithaca, NY 14850
(607) 255-5057
Specialties: current bird books

The Eyrie
Box 788
Redcliff, Alberta T0J 2P0
Specialties: birds of prey

Doug Kibbe
Box 34
Maryland, NY 12116
Specialties: current and out-of-print bird books

Patricia Ledlie
One Bean Road
PO Box 90
Buckfield, ME 04220
(207) 336-2778
Specialties: current and out-of-print bird books

Los Angeles Audubon Bookstore
7377 Santa Monica Boulevard
Los Angeles, CA 90046
(213) 876-0202
Specialties: current bird books

Melvin Marcher, Bookseller
6204 North Vermont
Oklahoma City, OK 73112
(405) 946-6270
Specialties: out-of-print ornithology and natural history

Massachusetts Audubon Bookstore
Great Road
Lincoln, MA 01773
(617) 259-9807
Specialties: current bird books

National Audubon Society
700 Broadway
New York, NY 10003
(800) 274-4201
Specialties: current bird books

Nature Press
40 Spruce Street
Columbus, OH 43215
(800) 532-6837
Specialties: Peterson field guides, Golden field guides, natural history

One Good Tern
1710 Fern Street
Alexandria, VA 22302
(800) 432-8376
Specialties: current bird books, field guides, finding guides

Peacock Books
Box 2024
Littleton, MA 01460
Specialties: out-of-print ornithology

Petersen Book Co.
Box 966
Davenport, IA 52805
(319) 355-7051
Specialties: current bird books

Rafiki Books
45 Rawson Avenue
Camden, ME 04843
(207) 236-4244
Specialties: Africa

Russ's Natural History
Books, Inc.
Box 1089
Lake Helen, FL 32744
(904) 228-3356
Specialties: field checklists,
field guides, natural history

Raymond M. Sutton, Jr.
430 Main Street
Williamsburg, KY 40769
(606) 549-3464
Specialties: current and out-of-
print bird books

Village Square Book Shoppe
Southwest 13th Street
Gainesville, FL 32608
(904) 336-7306

Specialties: birds and natural
history

Paul W. Weld
435 Lanning Road
Honeoye Falls, NY 14472
Specialties: current and out-of
print bird books

Birding Book Clubs

The Bird Book Source
Box 1088
Vineyard Haven, MA 02568
(800) 433-4811

The Natural Science Book Club
A Newbridge Book Club
3000 Cindel Drive
Delran, NJ 08075
(609) 786-9778

The Nature Book Society
PO Box 10875
Des Moines, IA 50336
(515) 284-3520

ORNITHOLOGY LIBRARIES

A number of ornithological libraries at universities, museums, and organizations are open to the public for research. This is an excellent way to look at hard-to-find or rare birding volumes. Since these libraries generally subscribe to all the scholarly birding journals, it is also a good way to look up interesting articles in back issues. In most cases, the books and journals about birds are part of a larger library of natural history. Contact the library well in advance to check on the hours and to arrange permission to visit. As a rule, permission to use the collection is granted if you have a serious area of research, but there may be a small fee and some restrictions. If you can't get to a distant library, speak to your local librarian about interlibrary loan programs.

UNITED STATES

California

California Academy
of Sciences Library
Golden Gate Park
San Francisco, CA 94118
(415) 750-7102

California State University
and College
Moss Landing Marine
Laboratories Library
Box 450
Moss Landing, CA 95039
(408) 755-8654

Hartnell Community College
O.P. Silliman
Memorial Library
156 Homestead Avenue
Salinas, CA 93901
(408) 755-6872

Natural History Museum of
Los Angeles County
Research Library
900 Exposition Boulevard
Los Angeles, CA 90007
(213) 744-3387

Point Reyes Bird
Observatory Library
4990 Shoreline Highway
Stinson Beach, CA 94970
(415) 868-1221

San Bernardino
County Museum
Wilson C. Hanna Library
2024 Orange Tree Lane
Redlands, CA 92373
(714) 798-8570

San Diego Society
of Natural History
Natural History
Museum Library
Box 1390
San Diego, CA 92112
(619) 232-3821

Colorado

Denver Museum of Natural
History Library
City Park
Denver, CO 80205
(303) 370-6361

United States Department
of Agriculture
Denver Wildlife Research
Center Library
Federal Center, Building 16
Box 25266
Denver, CO 80225
(303) 326-7873

Connecticut

Connecticut Audubon
Society Library
Fairfield Nature Center
2325 Burr Street
Fairfield, CT 06430
(203) 259-6305

Connecticut State Museum of
Natural History Library
University of Connecticut
U-23
Storrs, CT 06269
(203) 486-4460

Yale University
Ornithology Library
Peabody Museum
170 Whitney Avenue
New Haven, CT 06511
(203) 432-3797

Delaware

Delaware Museum of Natural
History Library
4840 Kennett Pike
Wilmington, DE 19807
(302) 658-9111

District of Columbia

Smithsonian Institution
Libraries
Special Collections Branch
MAH 5016
Washington, DC 20560
(202) 357-1568
Special Collection: Wetmore
Ornithology Collection

Florida

Archbold Biological
Station Library
Box 2057
Lake Placid, FL 33852
(813) 465-2571

Museum of Arts and Sciences
Bruce Everett Bates
Memorial Library
1040 Museum Boulevard
Daytona Beach, FL 32014
(904) 255-0285

United States National
Park Service
Everglades National Park
Reference Library
Box 279
Homestead, FL 33030
(305) 242-7800

Illinois

Chicago Academy of Sciences
Matthew Laughlin
Memorial Library
2001 North Clark Street
Chicago, IL 60614
(312) 549-0606

Field Museum of Natural
History Library
Roosevelt Road
and Lake Shore Drive
Chicago, IL 60605
(312) 922-9410
Special Collection: Ayer
Ornithology Library

Illinois State Museum of
Natural History and Art
Technical Library
Springfield, IL 62706
(217) 782-6623
Special Collection: R.M.
Barnes ornithology collection

University of Illinois
Biology Library
101 Burrill Hall
407 South Goodwin
Urbana, IL 61801
(217) 333-3654
Special Collection: Oberholser
reprint collection

Indiana

Earlham College
Joseph Moore Museum
Hadley Library
Box E-68
Richmond, IN 42374
(317) 983-1303

Kansas

University of Kansas
Spencer Research Library
Lawrence, KS 66045
(913) 864-4334
Special Collection: History of ornithology

Louisiana

Louisiana State University
Special Collections, Hill
Memorial Library
Baton Rouge, LA 70803
(504) 388-6551
Special Collection: E.A. McIlhenny Natural History collection

Maine

Maine Audubon Society
Environmental Library and
Teacher Resource Center
Box 6009
Falmouth, ME 04105
(207) 781-2330

Maryland

Baltimore Zoo
Arthur R. Watson Library
Druid Hill Park
Baltimore, MD 21217
(410) 396-6013

Natural History Society of
Maryland Library
2643 North Charles Street
Baltimore, MD 21218
(301) 235-6116

United States Fish and
Wildlife Service
Patuxent Wildlife Research
Center Library
Laurel, MD 20708
(301) 498-0235

Massachusetts

University of Massachusetts,
Amherst
Morrill Biological and
Geological Sciences Library
214 Morrill Science Center
Amherst, MA 01003
(413) 545-2674
Special Collection: Arthur Cleveland Bent ornithology collection

Manomet Bird
Observatory Library
Box 936
Manomet, MA 02345
(508) 224-6521

Michigan

Michigan State University
W.K. Kellogg
Biological Station
Walter F. Morofsky
Memorial Library
3700 East Gull Lake Drive
Hickory Corners, MI 49060
(616) 671-2310

Ornithology Libraries

University of Michigan
Natural Science Library
3140 Natural Science Building
Ann Arbor, MI 48109
(313) 764-1494

Wilson Ornithological Society
Josselyn Van Tyne
Memorial Library
Museum of Zoology,
University of Michigan
Ann Arbor, MI 48109
(313) 764-0457

Minnesota

Minnesota Department of
Natural Resources Library
500 Lafayette Road
St. Paul, MN 55155
(612) 297-4929

University of Minnesota
Bell Museum of Natural
History Library
10 Church Street, SE
Minneapolis, MN 55455
(612) 624-1639

Mississippi

Mississippi Museum of
Natural Science Library
The Fannye A. Cook Memorial
111 North Jefferson Street
Jackson, MS 39201
(601) 354-7303

Missouri

Burroughs-Audubon Society
Center and Library
RR 3, Box 120
Blue Springs, MO 64015
(816) 795-8177

New York

American Museum
of Natural History
Department of Library Services
Central Park West at 79th Street
New York, NY 10024
(212) 769-5400

Brooklyn Public Library
Grand Army Plaza
Brooklyn, NY 11238
(718) 780-7745

Cornell University
History of Science Collections
215 John M. Olin Library
Ithaca, NY 14853
(607) 255-4033
Special Collection: Hill
Collection of 18th- and 19th-
century North
American ornithology

Cornell University
Laboratory of
Ornithology Library
159 Sapsucker Woods Road
Ithaca, NY 14850
(607) 254-2403

Cornell University
Laboratory of Ornithology
Library of Natural Sounds
159 Sapsucker Woods Road
Ithaca, NY 14850
(607) 254-2404
Special Collection: 70,000
sound recordings

BOOKS, SOFTWARE, AND BEYOND

North Dakota

United States Fish and
Wildlife Service
Northern Prairie Wildlife
Research Center Library
Route 1, Box 96C
Jamestown, ND 58401
(701) 252-5363

Ohio

Cleveland Museum
of Natural History
Harold Terry Clark Library
One Wade Oval Drive
Cleveland, OH 44106
(216) 231-4600

Pennsylvania

Carnegie Museum of Natural
History Library
4400 Forbes Avenue
Pittsburgh, PA 15213
(412) 622-3264
Special Collections: W.E. Clyde
Todd Ornithological Reprint
Collection, John P. Robin
Library, G. Bernard Van Cleve
Library

Schuylkill Center for
Environmental Education
Library
8480 Hagy's Mill Road
Philadelphia, PA 19128
(215) 482-7300

Zoological Society of
Pennsylvania Library
3400 West Girard Avenue
Philadelphia, PA 19104
(215) 243-1100

Texas

Corpus Christi
Museum Library
1900 North Chaparral
Corpus Christi, TX 78401
(512) 883-2862

Dallas Museum of
Natural History
Mudge Rare Book Library
Box 150433
Dallas, TX 75315
(214) 670-8476

Rob and Bessie Welder Wildlife
Foundation Library
Drawer 1400
Sinton, TX 78387
(512) 364-2643
Special Collection: Quillen Egg
Collection

Washington

Seattle Public Library
Business and Technology
Department
1000 Fourth Avenue
Seattle, WA 98104
(206) 386-4634

Wisconsin

University of Wisconsin
at Madison
Zoological Museum Library
L.E. Noland Building
250 North Mills Street
Madison, WI 53706
(608) 262-3766

CANADA

Alberta

Environment Canada
Western and Northern
Region Library
4999 98th Avenue, Room 210
Edmonton, Alberta T6B 2X3
(403) 468-8951

British Columbia

Royal British Columbia
Museum Library
675 Belleville Street
Victoria, British Columbia
V8V 1X4
(604) 387-2916

Vancouver Public
Aquarium Library
Box 3232
Vancouver, British Columbia
V6B 3X8
(604) 685-3364

Manitoba

Delta Waterfowl and Wetland
Research Station
David Winton Bell
Memorial Library
RR 1
Portage La Prairie, Manitoba
R1N 3A1
(204) 239-1900

New Brunswick

Canadian Wildlife Service
Atlantic Region Library
Box 1590
Sackville, New Brunswick
E0A 3C0
(506) 536-3025

Ontario

Canadian Museum of Nature
Library and Archives
Box 3443, Station D
Ottawa, Ontario K1P 6P4
(613) 998-3923
Special Collection: R.M.
Anderson Collection

Canadian Wildlife Service
Ontario Region Library
49 Camelot Drive
Nepean, Ontario K1A 0H3
(613) 952-2406

Point Pelee National
Park Library
RR 1
Leamington, Ontario N8H 3V4
(519) 322-2365

Royal Botanical
Gardens Library
Box 399
Hamilton, Ontario L8N 3H8
(416) 527-1158

Quebec

Environment Canada
Quebec Region Library
1141 Route de l'Eglise,
7th Floor
Ste. Foy, Quebec G1V 4H5
(418) 649-6546

BOOKS, SOFTWARE, AND BEYOND

McGill University
Blacker/Wood Library
of Biology
Redpath Library Building
3459 McTavish Street
Montreal, Quebec H3A 1Y1
(514) 398-4744

Environment Canada,
Conservation and Protection
Canadian Wildlife Service
Quebec Region Library
1141 Route de l'Eglise
Box 10100
Ste. Foy, Quebec G1V 4H5
(418) 648-7062

Saskatchewan

Canadian Wildlife Service
Prairie Migratory Bird
Research Centre Library
115 Perimeter Road
Saskatoon, Saskatchewan
S7N 0X4
(306) 975-4087

ENGLAND

BirdLife International Library
32 Cambridge Road
Girton, Cambridge CB3 0PJ
(22) 327-7318

Natural History Museum
Department of Library Services
Cromwell Road
London SW7 5BD
(71) 938-9191

PERIODICALS

The periodicals listed below are magazines and newsletters of general birding interest. Some, such as *The Auk*, are scholarly journals oriented toward serious ornithologists. Although these journals can be quite technical, they are still of interest even to the nonscientist. Most large libraries receive a good assortment of general nature periodicals and at least some serious birding journals. The newsletters published by Audubon chapters and other birding organizations are another good source of interesting, accessible information. For information about these newsletters, see the section on organizations for birders.

Periodicals

American Birds
National Audubon Society
700 Broadway
New York, NY 10003
(212) 832-3200
Frequency: quarterly magazine

The Auk
American Ornithologists' Union
c/o Division of Birds
National Museum of Natural History
Washington, DC 20560
(202) 357-1496
Frequency: quarterly journal

Birder's Journal
Eight Midtown Drive, Suite 289
Oshawa, Ontario L1J 8L2
Frequency: bimonthly magazine

Birder's World
44 East Eighth Street, Suite 410
Holland, MI 49423
(616) 396-5618
Frequency: bimonthly magazine

Birding
American Birding Association
Box 6599
Colorado Springs, CO 80934
(800) 850-2473
Frequency: bimonthly journal

Birding World
Stonerunner, Cley-next-the-Sea
Holt, Norfolk NR25 7RZ
England
Frequency: monthly magazine

Bird Observer of Eastern Massachusetts
Box 236
Arlington, MA 02174
Frequency: bimonthly journal

Royal Society for the Protection of Birds
The Lodge
Sandy, Bedfordshire SG12 2DL
England
0767 680551

Bird Trends
Migratory Birds
Conservation Division
Canadian Wildlife Service
Ottawa, Ontario K1A 0H3
Frequency: annual magazine

BirdWatcher's Digest
Box 110
Marietta, OH 45750
(800) 421-9764
Frequency: bimonthly magazine

British Birds
Fountains, Park Lane
Blunham, Bedford MK44 3NJ
England
0767 40467
Frequency: monthly magazine

The BWD Skimmer
Box 110
Marietta, OH 45750
(800) 421-9764
Frequency: bimonthly newsletter
(800) 879-2473

BOOKS, SOFTWARE, AND BEYOND

The Canadian Field Naturalist
Francis R. Cook, Editor
RR 3
North Augusta, Ontario
K0G 1R0
Frequency: quarterly journal

Colonial Waterbirds
R. Michael Erwin, Editor
Patuxent Wildlife
Research Center
Laurel, MD 20708

The Condor
OSNA
Box 1897
Lawrence, KS 66044

Conservation Biology
Reed Noss, Editor
Department of Fisheries
and Wildlife
Oregon State University
Corvallis, OR 97331
Frequency: quarterly journal

Dutch Birding
Postbus 75611
1070 AP
Amsterdam, Netherlands
Frequency: monthly magazine

The Euphonic
Kurt Rademacher, Editor
Box 8045
Santa Monica, CA 93456
Frequency: quarterly journal

Florida Birding
Noel Wamer
502 East Georgia Street
Tallahassee, FL 32303
Frequency: bimonthly
newsletter

Ibis
The British
Ornithologists Union
c/o The Natural
History Museum
Sub-Department of
Ornithology
Tring, Herts HP23 6AP
England
Frequency: quarterly journal

Journal of Raptor Research
Carl D. Marti, Editor
Department of Zoology
Weber State University
Ogden, UT 84408
Frequency: quarterly journal

KBBW (Kachemak Bay Bird Watch)
Birchside Studio
Box 841
Homer, AK 99603
Frequency: quarterly newsletter

The Living Bird
Laboratory of Ornithology
Cornell University
159 Sapsucker Woods Road
Ithaca, NY 14850
(607) 254-BIRD
Frequency: quarterly magazine

Nature Society News
Purple Martin Junction
Griggsville, IL 62340
Frequency: monthly newsletter

Periodicals

North American Bird Bander
c/o Robert Pantle
35 Logan Hill Road
Candor, NY 13743
Frequency: quarterly journal

Partners in Flight
c/o Peter Stangel
National Fish and Wildlife
Foundation
1120 Connecticut Avenue NW,
Suite 900
Washington, DC 20036
Frequency: quarterly newsletter

Pennsylvania Birds
2469 Hammertown Road
PA
Frequency: quarterly journal

Refuge Reporter
Avocet Two
Avocet Crossing
Millwood, VA 22646
Frequency: biannual magazine

*The Russian Journal of
Ornithology*
c/o Eugene Potapow
Department of Zoology
South Parks Road
Oxford OX1 3PS
England

U.S. Birdwatch
U.S. Section Office, ICBP-US
c/o World Wildlife Fund
1250 24th Street NW
Washington, DC 20037
(202) 778-9563
Frequency: biannual magazine

Western Birds
c/o Philip Unitt, Editor
3411 Felton Street
San Diego, CA 92104
Frequency: quarterly journal

WildBird Magazine
Box 57900
Los Angeles, CA 90057
(213) 385-2222
Frequency: monthly

Wildlife Rehabilitation Today
Coconut Creek
Publishing Company
2201 NW 40th Terrace
Coconut Creek, FL 33066
Frequency: bimonthly
magazine

The Wilson Bulletin
Wilson Ornithological Society
OSNA
Box 1897
Lawrence, KS 66044
Frequency: quarterly journal

BIRD SLIDES

After actually going out into the field and seeing them, one of the best ways to learn about birds is to look at pictures of them. A favorite way to enjoy bird pictures is to look at color slides, especially at a meeting of your local bird club. Several sources of high-quality slides are listed below. All offer catalogs of inexpensive duplicate slides of hundreds of species. Remember that slides purchased from these sources are for noncommercial use only and may not be duplicated.

Cornell Laboratory of
Ornithology
Visual Services
159 Sapsucker Woods Road
Ithaca, NY 14850
(607) 254-2450

VIREO (Visual Resources
for Ornithology)
Academy of Natural Sciences
1900 Ben Franklin Parkway
Philadelphia, PA 19103
(212) 299-1069

Sea and Sage Audubon
Library of Nature Slides
Box 25
Santa Ana, CA 92702

BIRDING GEAR

One of the pleasures of birding is that it doesn't really require much in the way of equipment. A pair of binoculars, a field guide, sturdy shoes, and you're off birding in the field. But what's a hobby if it doesn't also have a lot of interesting paraphernalia? Happily, birding is no exception. Numerous manufacturers offer a wide variety of birding doodads, ranging from bird squeakers to video games, to say nothing of birding vests and birdfeeders. All sorts of useful birding stuff, along with a good selection of books, field guides, and tapes, can usually be found at your local nature store, shops run by nature organizations, or through a number of outstanding mail-order sources.

There are far too many nature stores to list them all individually here. The better-known retail outlets catering to birders are given, but the many, many fine stores offering primarily birdfeeders, seed, and other supplies for backyard birding are not listed. The list does include the headquarters of the larger franchise operations—contact the head office for the location of the store nearest you. Bear in mind that many retail stores also have mail-order catalogs. These make excellent reading when you can't be birding.

American Birding
Association Sales
Box 6599
Colorado Springs, CO 80934
(800) 634-7736

BWD Direct
Box 110
Marietta, OH 45750
(800) 879-2473

The Chickadee Nature Store
1330-L Wirt
Houston, TX 77055
(713) 956-2670

The Crow's Nest Birding Shop
Cornell Laboratory
of Ornithology
159 Sapsucker Woods Road
Ithaca, NY 14850
(607) 254 2400

The Nature Company
(franchised stores)
750 Hearst Avenue
Berkley, CA 94710
(510) 644-1337

Southern Naturalist
Tallahassee, FL
(800) 578-0879

Three Coast Emporium
Box 30482
Charleston, SC 29417

Wild Bird Center
(franchised stores)
7687 McArthur Boulevard
Cabin John, MD 20818
(800) WILDBIRD

Wild Bird Marketplace
(franchised stores)
710 West Main Street
New Holland, PA 17557
(717) 354-2841

Wild Birds Unlimited
(franchised stores)
3003 East 96th Street, Suite 201
Indianapolis, IN 46240
(317) 571-7100

Wildlife Company
46 Monroe Street
Ellicottville, NY 14731
(716) 699-5195

Wood Creek
3018 Marshall Avenue
Cincinnati, OH 45220
(513) 684-0400

BIRDING ON-LINE

Software and on-line services are rapidly growing areas of birding. Listing software has been around for a while and has become increasingly cheaper, more sophisticated, and easier to use. Contact the manufacturers for program details. Most will send a sample disk at no or nominal charge. More recently, software programs that combine sight and sound have become available, chiefly in the form of birdsong study programs.

On-line birding bulletin boards have become a very popular way to "talk" with birders across the country and even around the world and to keep tabs on rare-bird alerts. On-line databases and network listservers provide information, alerts, and conversation.

The information here is as accurate and complete as possible, but in this fast-moving field, products and new services constantly arise.

Listing Programs

Aveformes
Golias Software
Box 3190
Johnstown, PA 15904
Requirements: IBM-compatible with hard disk and 640K memory

AviSys versions 2.0 and 3.0
Perceptive Systems
PO Box 3530
Silverdale, WA 98383
(800) 354-7755
Requirements: IBM-compatible with hard drive and 640K memory

BirdBase 2
Santa Barbara Software Products
1400 Dover Road
Santa Barbara, CA 93013
(805) 963-4886
Requirements: IBM-compatible with hard drive and 640K memory

Bird Brain 2.0 Birding Database
Ideaform Inc.
Box 1540
Fairfield, IA 52556
(515) 472-7256
(800) 779-7256
Requirements: Macintosh with system 6

BirdCount+
International Innovations Inc.
61 Maple Avenue
White Plains, NY 10601
(914) 686-1218

Requirements: IBM-compatible with hard drive and 640K memory

The Birder's Notebook
GG Software Co., Suite 209
1266 Furnace Brook Parkway
Quincy, MA 02169
(800) 435-4460
Requirements: IBM-compatible

Datahawk Version 2
Turnstone Software
1838 Barry Avenue, Suite 12
Los Angeles, CA 90025
(800) 654-5676
Requirements: IBM-compatible with hard drive

Flexi-List Version 2.1
Parkway Software
Box 275
Villanova, PA 19085
Requirements: IBM-compatible, 256K memory

KwikBirdlist
Cliff Lamere
15 Saradale Avenue
Albany, NY 12211
(518) 462-9827
Requirements: IBM-compatible with WordPerfect 5.1

Macmerlin
Wholelife Systems
Box 162
Rehoboth, NM 87322
(505) 863-4751
Requirements: Macintosh with hypercard

MacPeregrine
WholeLife Systems
Box 162
Rehoboth, NM 87322
(505) 863-4751
Requirement: Macintosh

Plover
Sandpiper Software
Nine Goldfinch Court
Novato, CA 94947
Requirements: IBM-compatible

Sialis
Alfred Milch
461 Palmer Avenue
Teaneck, NJ 07666
Requirements: IBM-compatible

Bird Song Programs

Bird Song Master
Micro Wizard
(614) 846-1077
Requirements: IBM-compatible with CD-ROM drive

Bird Sounds Tutor
NetWings Co.
3936 Pine Tree Road
Quincy, IL 62301
(217) 223-6905
Requirements: IBM-compatible with hard disk and sound board

Other Software

BIRDLEXI
Santa Barbara
Software Products
1400 Dover Road

Santa Barbara, CA 93013
(805) 963-4886
Function: spell-checker for ornithology
Requirements: IBM-compatible with hard drive and 640K memory

Life Cycles
Reflections of Nature
W14437 Hookers Road
Gilman, WI 54433
(715) 668-5730
Function: phenology program to track cyclic natural history events
Requirements: IBM-compatible

Snipe Hunt
Brown Bag Software
Box 18796
Cleveland Heights, OH 44118
Function: introduction to birding for kids
Requirements: IBM-compatible with Windows and sound support

Bulletin Boards

Birdwatcher's Nest Echo
UA Today, message area 12
(602) 629-0502

Cocoino County Bulletin Board
(415) 861-8290

Osprey's Nest Bulletin Board for Birders
(301) 989-9036

The Pacific Rim
(619) 278-7361

QUAKE
(818) 362-6092

Southern Arizona Birding Bulletin Board
(602) 881-4280

Network Listservers

National Birding Hotline Cooperative (NBHC)

BIRDEAST

BIRDCNTR

BIRDWEST

BIRDCHAT

Organizations

Newburyport Birders' Exchange
Eight Columbia Way
Plum Island, MA 01951
Available on: Bitnet/Internet, Usenet, other e-mail services
Log on: LISTSERV @ ARIZVM1
Listings: BIRDCHAT, BIRDEAST, BIRDCNTR, BIRDWEST

APPENDIX A

AMERICAN BIRDING ASSOCIATION CODE OF ETHICS

I. Birders must always act in ways that do not endanger the welfare of birds or other wildlife.

- Observe and photograph birds without knowingly disturbing them in any significant way.
- Avoid chasing or repeatedly flushing birds.
- Only sparingly use recordings and similar methods of attracting birds and do not use these methods in heavily birded areas.
- Keep an appropriate distance from nests and nesting colonies so as not to disturb them or expose them to danger.
- Refrain from handling birds or eggs unless engaged in recognized research activities.

II. Birders must always act in ways that do not harm the natural environment.

- Stay on existing roads, trails, and pathways whenever possible to avoid trampling or otherwise disturbing fragile habitat.
- Leave all habitat as it was found.

III. Birders must always respect the rights of others.

- Observe all laws and the rules and regulations which govern public use of birding areas.

- Practice common courtesy in our contacts with others. For example, limit requests for information, and make them at reasonable hours of the day.

- Always behave in a manner that will enhance the image of the birding community in the eyes of the public.

IV. Birders in groups should assume special responsibilities.

As group members, we will

- Take special care to alleviate the problems and disturbances that are multiplied when more people are present.

- Act in consideration of the group's interest, as well as our own.

- Support by our actions the responsibility of the group leader(s) for the conduct of the group.

As group leaders, we will

- Assume responsibility for the conduct of the group.

- Learn and inform the group of any special rules, regulations, or conduct applicable to the area or habitat being visited.

- Limit groups to a size that does not threaten the environment or the peace and tranquility of others.

- Teach others birding ethics by our words and example.

Reprinted by permission of the American Birding Association

APPENDIX B

BIRDING HOTLINES

Many National Audubon Society local chapters, local Audubon societies, and other birding organizations have birding hotlines—recorded messages announcing general birding conditions in the region and details on where to find any rare and unusual birds that have been reported. The messages are changed at least once a week and often more frequently. The information is often quite detailed; have pen and paper handy to note directions and phone numbers. The expense to you is only the cost of the phone call. Hotline numbers tend to change fairly often, usually as the answering machine moves from one volunteer's home to another. Whoever answers at the other end can usually tell you the new number.

The North American Rare Bird Alert (NARBA) is a fee-based service sponsored by the Houston Audubon Society. It provides information about sightings of unusual birds throughout North America, not just in a particular state or region. The number to hear the recorded message is given only to subscribers. Funds from NARBA subscriptions go to support an extensive system of refuges on the upper Texas coast. For more information, call (800) 458-BIRD.

UNITED STATES

Alabama

Statewide: (205) 987-2730

Alaska

Statewide: (907) 338-2473

Arizona

Phoenix: (602) 832-8745
Tucson: (602) 798-1005

Arkansas

Statewide: (501) 753-5853

APPENDIX B

California

Arcata: (707) 826-7031

Los Angeles: (213) 874-1318

Monterey: (408) 375-9122

Morro Bay: (805) 528-7182

Northern California:
(510) 524-5592

Orange County: (714) 563-6516

Sacramento: (916) 481-0118

San Bernadino: (909) 793-5599

San Diego: (619) 479-3400

Santa Barbara: (805) 964-8240

Southwest Sierra/San Joaquin:
(209) 782-1237

Colorado

Statewide: (303) 279-3076

Connecticut

Statewide: (203) 254-3665

Delaware

Statewide: (215) 567-2473

District of Columbia

Districtwide: (301) 652-1088

Florida

Statewide: (813) 984-4444
Miami: (305) 667-7337
Lower Keys: (305) 294-3438

Georgia

Statewide: (404) 509-0204

Idaho

Southeast: (208) 236-3337

Illinois

Central: (217) 785-1083
Chicago: (708) 671-1522

Indiana

Statewide: (317) 259-0911

Iowa

Statewide: (319) 338-9881
Sioux City: (712) 262-5958

Kansas

Statewide: (913) 372-5499
Kansas City: (913) 342-2473

Kentucky

Statewide: (502) 894-9538

Louisiana

Baton Rouge: (504) 293-2473
New Orleans: (504) 246-2473

Maine

Statewide: (207) 781-2332

Maryland

Statewide: (301) 652-1088

Massachusetts

Boston: (617) 259-8805
Western Region:
(413) 253-2218

Birding Hotlines

Michigan

Statewide: (616) 471-4919
Detroit: (313) 477-1360
Sault Ste. Marie: (705) 256-2790

Minnesota

Statewide: (612) 827-3161
Duluth: (218) 525-5952

Mississippi

Coast: (601) 467-9500

Missouri

Statewide: (314) 445-9115
Kansas City: (913) 342-2473
St. Louis: (314) 935-8432

Montana

Statewide: (406) 626-2473

Nebraska

Statewide: (402) 292-5325

Nevada

Statewide: (702) 649-1516
Northwestern Region:
(702) 324-2473

New Hampshire

Statewide: (603) 224-9900

New Jersey

Statewide: (908) 766-2661
Cape May: (609) 884-2626

New Mexico

Statewide: (505) 662-2101

New York

Albany: (518) 439-8080
Buffalo: (716) 896-1271
Cayuga Lakes Basin:
(607) 254-2429
Lower Hudson Valley: (914) 666-6614
New York City: (212) 979-3070
Rochester: (716) 461-9593
Syracuse: (315) 682-7039

North Carolina

Statewide: (704) 332-2473

Ohio

Blendon Woods Metro Park:
(614) 895-6222
Cincinnati: (513) 521-2847
Cleveland: (216) 321-7245
Columbus: (614) 221-9736
Northwest Region:
(419) 875-6889
Southwest Region:
(513) 277-6446
Youngstown: (216) 742-6661

Oklahoma

Oklahoma City: (405) 373-4531

Oregon

Statewide: (503) 292-0661

Pennsylvania

Allentown: (215) 759-5754
Philadelphia: (215) 567-2473
Southeast/South-Central Region: (215) 383-8840
Western Region: (412) 963-0560
Wilkes-Barre: (717) 825-2473

Rhode Island

Statewide: (401) 231-5728

South Carolina

Statewide: (704) 332-2473

Tennessee

Statewide: (615) 356-7636
Chattanooga: (615) 843-2822

Texas

Statewide: (713) 992-2757
Austin: (210) 483-0952
Lower Rio Grande Valley: (210) 565-6773
Northcentral Region: (817) 261-6792
San Antonio: (210) 733-8306
Sinton: (210) 364-3634

Utah

Statewide: (801) 538-4730

Vermont

Statewide: (802) 457-4861

Virginia

Statewide: (804) 238-2713 and (301) 652-1088

Washington

Statewide: (206) 526-8266

Wisconsin

Statewide: (414) 352-3857
Madison: (608) 255-2476

Wyoming

Statewide: (307) 265-2473

CANADA

Alberta

Calgary: (403) 237-8821

British Columbia

Vancouver: (604) 737-9910
Victoria: (604) 592-3381

New Brunswick

Provincewide: (506) 382-3825

Nova Scotia

Provincewide: (902) 852-2428

Ontario

Provincewide: (519) 586-3959
Durham: (905) 428-9972
Hamilton: (905) 648-9537
Long Point Bird Observatory: (519) 586-3959

Ottawa: (613) 761-1967
Sault Ste. Marie: (705) 256-2790
Toronto:
(416) 350-3000 ext. 2293
Windsor/Detroit:
(313) 477-1360
Windsor/Pt. Pelee:
(519) 252-2473

Quebec

Montreal: (514) 355-7255
(in French); (514) 355-6549
(in English)

Quebec City Region
(in French): (418) 660-9089
Eastern Region (in French):
(819) 778-0737
Western Region (in French):
(819) 563-6603
Bas St. Laurent (in French):
(418) 725-5118
Sagueny/Lac St. Jean (in
French): (418) 696-1868

Saskatchewan

Regina: (306) 761-2094

APPENDIX C

STATE BIRDS

State	Mascot bird
Alabama	northern flicker (yellowhammer)
Alaska	willow ptarmigan
Arizona	cactus wren
Arkansas	northern mockingbird
California	California quail
Colorado	lark bunting
Connecticut	American robin
Delaware	blue hen chicken
District of Columbia	wood thrush
Florida	northern mockingbird
Georgia	brown thrasher
Hawaii	nene
Idaho	mountain bluebird
Illinois	northern cardinal
Indiana	northern cardinal
Iowa	American goldfinch
Kansas	western meadowlark
Kentucky	northern cardinal
Louisiana	brown pelican
Maine	black-capped chickadee
Maryland	northern ("Baltimore") oriole
Massachusetts	black-capped chickadee
Michigan	American robin
Minnesota	common loon
Mississippi	northern mockingbird
Missouri	eastern bluebird

State Birds

State	*Mascot bird*
Montana	western meadowlark
Nebraska	western meadowlark
Nevada	mountain bluebird
New Hampshire	purple finch
New Jersey	American goldfinch
New Mexico	greater roadrunner
New York	eastern bluebird
North Carolina	northern cardinal
North Dakota	western meadowlark
Ohio	northern cardinal
Oklahoma	scissor-tailed flycatcher
Oregon	western meadowlark
Pennsylvania	ruffed grouse
Rhode Island	Rhode Island red chicken
South Carolina	Carolina wren
South Dakota	ring-necked pheasant
Tennessee	northern mockingbird
Texas	northern mockingbird
Utah	California gull
Vermont	hermit thrush
Virginia	northern cardinal
Washington	American goldfinch
West Virginia	northern cardinal
Wisconsin	American robin
Wyoming	western meadowlark

NOTES AND UPDATES

NOTES AND UPDATES

Notes and Updates

NOTES AND UPDATES

NOTES AND UPDATES